Sustainable Development Goals Series

The **Sustainable Development Goals Series** is Springer Nature's inaugural cross-imprint book series that addresses and supports the United Nations' seventeen Sustainable Development Goals. The series fosters comprehensive research focused on these global targets and endeavours to address some of society's greatest grand challenges. The SDGs are inherently multidisciplinary, and they bring people working across different fields together and working towards a common goal. In this spirit, the Sustainable Development Goals series is the first at Springer Nature to publish books under both the Springer and Palgrave Macmillan imprints, bringing the strengths of our imprints together.

The Sustainable Development Goals Series is organized into eighteen subseries: one subseries based around each of the seventeen respective Sustainable Development Goals, and an eighteenth subseries, "Connecting the Goals," which serves as a home for volumes addressing multiple goals or studying the SDGs as a whole. Each subseries is guided by an expert Subseries Advisor with years or decades of experience studying and addressing core components of their respective Goal.

The SDG Series has a remit as broad as the SDGs themselves, and contributions are welcome from scientists, academics, policymakers, and researchers working in fields related to any of the seventeen goals. If you are interested in contributing a monograph or curated volume to the series, please contact the Publishers: Zachary Romano [Springer; zachary.romano@springer.com] and Rachael Ballard [Palgrave Macmillan; rachael.ballard@palgrave.com].

More information about this series at http://www.springer.com/series/15486

Ali Cheshmehzangi

Urban Health, Sustainability, and Peace in the Day the World Stopped

Springer

Ali Cheshmehzangi
Department of Architecture
and Built Environment
University of Nottingham Ningbo China
Ningbo, Zhejiang, China

ISSN 2523-3084 ISSN 2523-3092 (electronic)
Sustainable Development Goals Series
ISBN 978-981-16-4887-8 ISBN 978-981-16-4888-5 (eBook)
https://doi.org/10.1007/978-981-16-4888-5

This Springer imprint is published by the registered company Springer Nature Singapore Pte Ltd.
The registered company address is: 152 Beach Road, #21-01/04 Gateway East, Singapore 189721, Singapore

Acrimonious Production

Unjustified Inequalities

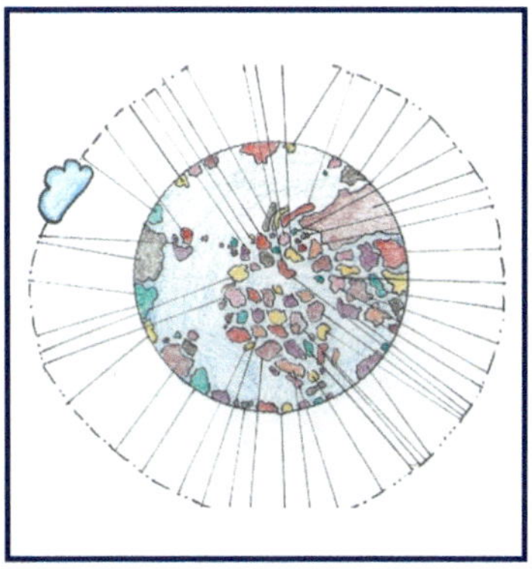

Segregated Connectivity

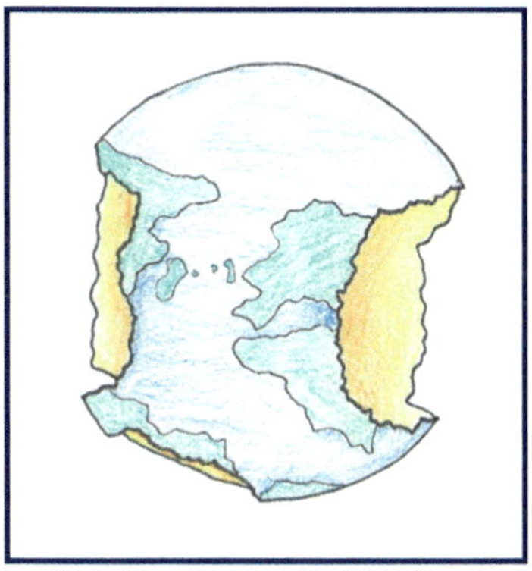

Eating-up Resources

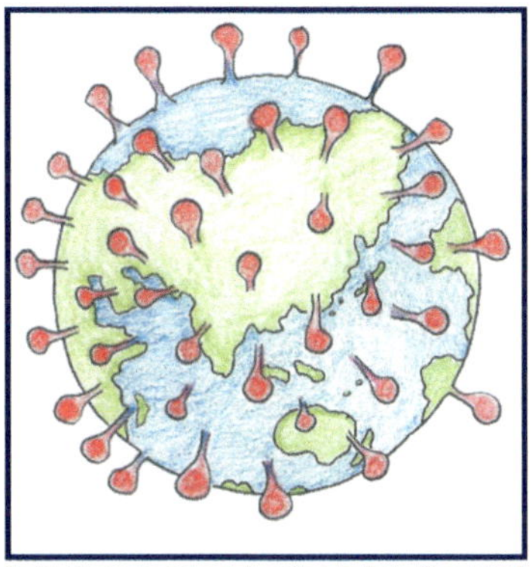

Earthy Pandemic

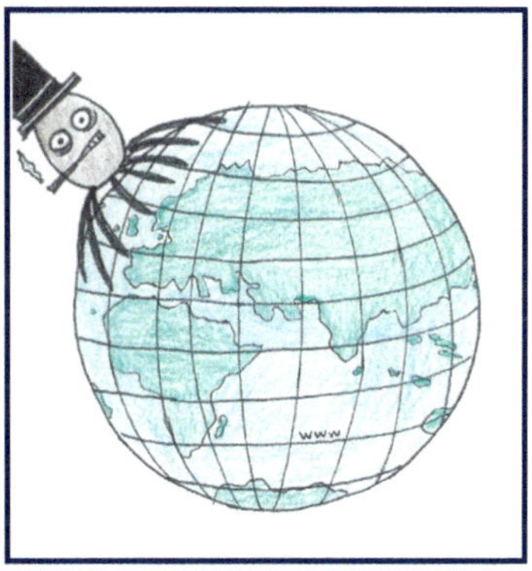

The World-Wide-Web

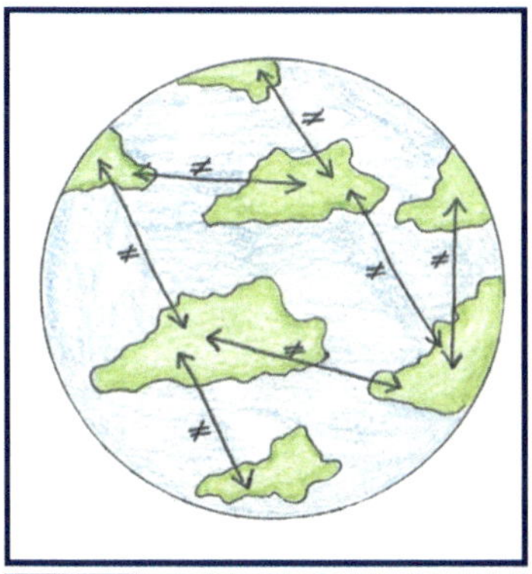

Unjust Globalisation

Pseudo-Green Prosperity

Unhealthy Prospect …

Images drawn by Ali Cheshmehzangi

To ***Mr. Maziyar Ordoobadi****, a good old friend who had a very short life. His early departure was a sudden tragedy for those who knew him. He was a true-hearted young lad who just liked the shades of clouds making colours on mountains, on a partly cloudy day. He had a pure heart, he offered peace, and he was vulnerable. His resilience was admirable, and his weakness was the main challenge. His sincere attitude reminds me of a peaceful time, a cold look at his palm, a moment of pause.*
I remember I had to tell him a white lie to give him hope and keep his peace. His health was already deteriorating, and our ephemeral friendship was not sustainable.

He left us 22 years ago.

I also dedicate this book to those believers, activists, and innovators who continuously seek peace against the corrupted cycles of everyday life. I genuinely hope our peace-makers could remain as peace-keepers, too.

Acknowledgements

This book is an early outcome of the project funded by the Ministry of Education, Culture, Sports, Science, and Technology (MEXT), Japan, and Hiroshima University, Japan. I thank the funders and the university for their support in giving me the opportunity to study peace and sustainability. While I wish I could have conducted more workshops in Hiroshima during this time, I hope I get the opportunity to further extend my collaboration with them after the ongoing pandemic. In this recently funded programme, we study ICT-mediated platforms for smart-resilient cities and communities. Part of the work is focused on evaluating positive peace and urban health. I also acknowledge the National Natural Science Foundation of China (NSFC) for funding project numbers 71850410544 and 71950410760, used for material purchase and recruitment of local research interns in China.

Lastly, I acknowledge those who give and take away peace from us. Both groups are equally important. When we have peace, we can see the values it offers us; when we do not, we value it more. All we have to say is the following for those who should remember well. "There is absolutely no neglect apprehended; there are no guarantees. Life opposes viciousness endlessly, yet our unity faces other retaliation; equilibrium verifies, & explications revive. Otherwise, universal reactions connect obedience, neglecting never-ending eruptive concerns, & the illusions of nature. Interestingly, sabotaging battles exist & yield our neutral destiny. This harmonisation is special for unified circumstances, knowledge, intelligence, & nonstop growth, where our reactions long divergence. Having optimal prosperity & equity yearn outstanding ubiquitous kindness, not opting wantedness in treasures…May your lives obey valued existence, may your best extracts amity & unlocking thriving years".

HIROSHIMA

www.nerps.org

University of Nottingham
UK | CHINA | MALAYSIA

www.nottingham.edu.cn

About This Book

Urban Health, Sustainability, and Peace is the first book that attempts to put these three critical areas together. This novelty approaches the subject matter by delving into evaluating what works, what does not work, and what should be done to achieve healthy cities. We believe this book will be beneficial to a wide range of stakeholders, particularly policymakers, planners, and developers, who continuously shape and reshape the structure and environments of our cities and communities. Unfortunately, in most cases, the healthiness of the cities may not be of their immediate concern. Nevertheless, it is the concern of the end-users, citizens, or simply those who live and work in cities and communities worldwide. To safeguard peace in cities, one has to consider sustaining urban health; that is the main aim of this book.

The ongoing pandemic gives us an excellent reason to study cities' health. During such a disruptive time, we detect many flaws in cities and communities around the world. We primarily identify the negative impacts on sustainability and peace in cities. In order to sustain a healthy city, this book evaluates six sustainability dimensions of physical, environmental, economic, social, institutional, and technical. It then utilises eight primary dimensions of positive peace, evaluating critical areas for future considerations in urbanism. These considerations include making cities smarter, more resilient, and more sustainable. The book's ultimate goal is to highlight how we should progress to maintain and sustain urban health.

As a continuation to *The City in Need*, this book covers the nexus between urban health, sustainability, and peace. Furthermore, by reflecting on the ongoing pandemic crisis, metaphorically labelled as "The Day the World Stopped", we delve into some key areas beyond the usual planning and policy guidelines. Lastly, the book intends to highlight what has not been studied before, i.e. the relationship between urban health, sustainability, and peace.

Prologue

The most essential thing in life is peace, and yet, we continuously and falsely seek everything else. We choose our wrong quests, we seek for the wrong, and we chase the wrong. All without knowing how to find peace, without knowing how to keep it, and without knowing how to cherish it properly. With peace, there is continuity. But it may not be sustainable, and it may not be healthy. We have to seek more. We have to aim higher—much higher than this.

I always believed the two sides of humanity's spectrum are "peace" and "corruption". I am confident we are currently far from the former. We are so tangled with our incessant progress and growth, without questioning the limits, not reaching the stability that is needed, and continuously sabotaging the balance that is vital to our existence. Given the significance of the COVID-19 pandemic impact on our society, we cannot just move forward without questioning the range of deficiencies, inactions, incapability, corruption, and inequalities around us. Our unceasing silence is an injudicious crime, feeding the violence against the peace we desire. We ought to reflect on the situation more considerably and responsibly to know what worked and what did not work during this almost never-ending pandemic. Nonetheless, this particular event is just a test for us to realise the depth of corruption and greed that have made us focus on continuous growth and unsustainable progressions. The harms we have caused become larger stones in our path towards a healthy and peaceful sustainable development. These have just happened, and we are at fault since the inception.

The bottleneck to our progress to achieve sustainability and sustainable development is ourselves, "the human beings". There is a lack of collective and global push, scarcity of political will, and many other insufficiencies representing our voracity and egotism in our behaviours. Our unsustainable preferences dictate us continuously, and we choose to select destructive behaviours. Such behaviours eventually become cravings, values, and patterns of our disregarding actions. The discarded minds cause adversities, not knowing that our values keep changing for the worse, and our goals and targets have no clear end to ensure stability and peace for all. Our over-reliance on technologies is growing fast in the so-called digital age, at a time where our neglect would lead to further inequalities, injustice, and imbalance. The latter is of our concern, i.e. the lack of balance between human and the environment, which is simply caused by us for years, decades, and even centuries of neglect, inactions, misbehaviours, and short memories of the past. Our history keeps repeating itself as we fail to learn and reflect on our preceding failures. We fail to become better as we feed our corruption and greed, as we think of everlasting growth, and we (often) do not see the end with a conscious mind. The harms we continue to cause are the ones that hit us back throughout this so-called progression. Yet, we get further away from humanity and the balance that is requisite of having healthier living environments.

Over the years, we have engineered everything. Not only so, but also that we have engineered everything for ourselves, pinpointed around our needs and demand—an endless enigma to our not-so-looked-after cradle. This paradox is not necessarily for the best. Our shift from ecoism to egoism concerns very little of the other beings of our ecosystem, without whom our decline is inevitable. Our investments are wrongly made, and our harmonious living is disrupted. Thus, we see growing challenges and some that we are yet to witness and experience. I do not suggest acts of entreating for celestial miracles nor do I suggest believing in such hysterical magic. Instead, I advocate nurturing humanity beyond just being humans. I recommend putting health with sustainability and peace and towards a more all-inclusive future. Our take on this ought to bring back the health and well-being of our environments, as we must become more considerate about our surroundings. After all, this is our colossal home, and we are not necessarily alone. We have to redefine our sustainability plans and agendas and start with game-changing ideas, more than just the already many reports of neglect. Lastly, we have to nurture peace as it could cultivate better humanity in future. Alas, we claim to be civilised, but we just pretend to be.

As we progress, we have a collective vision of a better future—but at what price? Besides, how do we define a "better future"? How do we create orderly progress that could sustain in a healthy way? And what values and targets shape our goal of a better future? As we continue to progress, we note many disorders, discrepancies, and distortions. Our minds are filled with greed and agony, a sort of inevitable chase that suggests an unhealthy development cycle. This leads to our egoism, which keeps us away from our inner peace and the outer peace that we could (possibly) offer. We believe in superficial pathways, weakening our minds for insecurity and praising the wrong. Then, our procrastination to make the right step forward becomes the cause of selecting the wrong choices, making immoral decisions, and focusing on false aspiration. All these together may just keep us away from the peace that we continuously seek—and we hope we could wake the mass by at least finding the right path(s) ahead. I believe this may just happen, but the question is when?

We trust this book is a good starting point to bring together the three key themes of health, sustainability, and peace. This novelty is (hopefully) a foundation for future research that must entail revisions, reforms, restructuring, and reconsideration in and of many things. The book holistically explores six dimensions of physical, environmental, economic, social, institutional, and technical, an approach that has been rarely seen in academia and scholarly work. Thus, we are rest assured the book could serve as a comprehensive example of scrutinising the health–sustainability–peace (HSP) nexus in cities and communities. We believe this direction is more meaningful than those recurrent rusty strategies, which have (so far) failed to advocate peace and prosperity in our society. The reason for this book is to expand on HSP research, develop new thinking, and help us reconsider the effectiveness of earlier agendas, guidelines, and strategies. After all, to achieve healthier and more sustainable living environments, we ought to put peace in the heart of our progress and development. If we fail to do so, we risk our humanity. If we succeed, we know that we have learned enough from the day the world stopped.

After many years of studying sustainability and sustainable development, I realised the missing part, i.e. peace. Not too late, but I realised this now. I now can verify that peace is more important than sustainability, and everything else is pinpointed around it. Sustainability alone is meaningless, and with peace-making thinking, we can complete it for the best. Unfortunately, most of our sustainability agendas either do not target peace or vaguely include it as one of the targets, goals, or indicators. However, now that I studied the HSP nexus in more detail, I realised that sustainability is mainly a greenwashing ideology without considering peace. Now I can confirm that, without peace, we are just deceiving ourselves of an ideal future far from reach. We cannot continue to manipulate our minds without having peace at the heart of sustainability and sustainable development progression. After all, it is only in a peaceful environment that one can achieve sustainability, or else, we are chasing the wrong thing. If we do so only in a peaceful environment, we will certainly reach health and (can) sustain it for a better future. So, let us find inner peace and offer a better outer peace for everybody and everything else. Let us learn from this day before it is late, and let us learn from the day the world stopped.

May 2021

Prof. Ali Cheshmehzangi
Visiting Professor, NERPS, Hiroshima University (HU)
Hiroshima, Japan

Head of Department and Professor
University of Nottingham Ningbo China (UNNC)
Ningbo, China

Director of Urban Innovation Lab

Contents

About the Author

Ali Cheshmehzangi is Visiting Professor at Network for Education and Research on Peace and Sustainability (NERPS), Hiroshima University, Hiroshima, Japan. He is Professor of Architecture and Urban Design at the University of Nottingham Ningbo China. He holds a Ph.D. degree in Architecture and Urban Design, a master's degree in Urban Design, a graduate certificate in Professional Studies in Architecture, and a bachelor degree in Architecture. He is an urbanist and urban designer by profession and by heart. He studies cities and city transitions, sustainable urbanism, and integrated urban design strategies.

He is Head of the Department of Architecture and Built Environment, Director of the Centre for Sustainable Energy Technologies (CSET), and Director of Urban Innovation Lab at the University of Nottingham Ningbo China. More recently, he works on two research projects on "Integrated Urban Modelling" and "ICT-based smart technologies for resilient cities". Some of his previous projects are: "smart eco-cities in China and Europe", "low-carbon town planning in China", "green infrastructure of cities", "nature-based solutions in China", "toolkit for resilient cities", "sponge city programme" and "green development in China", "low carbon and climate-resilient planning", and other urban transition studies.

So far, he has +110 published journal papers and eight other published books. His books are titled *Designing Cooler Cities* (2017—with Chris Butters), the award-winning *Eco-development in China* (2018—with Wu Deng), *Sustainable Urban Development in the Age of Climate Change* (2019—with Ayotunde Dawodu), *Identity of Cities and City of Identities* (2020), *The City in Need* (2020), *Urban Memory in City Transitions* (2021), *Sustainable*

Urbanism in China (2021—with Ayotunde Dawodu and Ayyoob Sharifi), and *China's Sustainability Transitions* (2021—with Hengcai Chen). His forthcoming book is an edited volume, titled *Green Infrastructure in Chinese Cities*.

1 Urban Health During the Pandemic: Why Does It Matter?

Our extreme divisions may lead to catastrophic events.
—Ali Cheshmehzangi, 2017

1.1 A General Overview of Urban Health in the Day the World Stopped

In late 2019, the world was still far from the adversities of the novel coronavirus disease. Soon after, the unknown disease grasped the opportunity to create the most significant disruption of our time. This led to a once-in-a-century disease outbreak, which was then eventually announced pandemic on the 11 March 2020. The pandemic came across as an unprecedented hidden calamity, which is still ongoing, progressing, and disrupting the so-called normal day-to-day activities. It simply put the world on halt, and hence, we hope to reflect on the situation more carefully. This reflection helps us highlight some of the lessons learnt and suggest what should be done afterward. Such experience makes us believe we have much to learn, and we have no doubts about solutions that could lead us to a more sustainable future.

"*Urban Health, Sustainability, and Peace*" is the first book that attempts to put these three critical areas together—i.e. health, sustainability, and peace. This novelty approaches the subject matter by delving into evaluating what works, what does not work, and what should be done to achieve healthy cities. We believe this book is beneficial to a wide range of stakeholders, particularly policymakers, planners, and developers, who continuously shape and reshape the structure and environments of our cities and communities. Unfortunately, in most cases, the healthiness of the cities may not be of their immediate concern. Nevertheless, it is the concern of the end-users, citizens, or simply those who live and work in cities and communities worldwide. To safeguard peace in cities, one has to consider sustaining urban health, and that is the main aim of this book.

The ongoing pandemic gives us an excellent reason (if not just a humble excuse) to study cities' health. During such a disruptive time, we detect many flaws in cities and communities around the world. These could be defined as deficiencies, shortages, shortfalls, mismanagement, mistreatments, etc., which usually lead to lack of frameworks, poor quality governance, deprivation, inequalities, inequity, corruption, and injustice. In this book, we primarily identify the negative impacts on sustainability and peace in cities, especially those that were caused by the pandemic. In order to sustain a healthy city, we explore multiple dimensions, which are studied individually and are often practiced together.

A. Cheshmehzangi, *Urban Health, Sustainability, and Peace in the Day the World Stopped*,
Sustainable Development Goals Series, https://doi.org/10.1007/978-981-16-4888-5_1

Thus, this book evaluates six sustainability dimensions: physical, environmental, economic, social, institutional, and technical. It then utilises eight primary dimensions of positive peace (see Chap. 2) to evaluate critical areas for future considerations in urbanism. These considerations include making cities smarter, more resilient, and more sustainable. We hope citizens/residents also follow the same trends, just to become smarter, more resilient, and more sustainable, too.

As a continuation to the earlier book on "The City in Need" (Cheshmehzangi 2020a), this book covers the nexus between urban health, sustainability, and peace. This nexus is identified as the HSP nexus for cities and communities. We note that this is the first time in scholarly research that these three overarching themes come together. By reflecting on the ongoing pandemic crisis, metaphorically labelled as "The Day the World Stopped", we delve into some key areas beyond the usual planning and policy guidelines. Lastly, the book intends to highlight what has not been studied before, i.e. the relationship between urban health, sustainability, and peace.

To start, we explore more of the urban health issues and delve into further arguments on disruptions and their impacts on society. Throughout many aspects that would be covered in the book, we also highlight some opportunities and advancements, particularly in facing or dealing with the pandemic. Our common sense would be to separate these, but we hope the book narrative brings them together, i.e. identifying opportunities against disruptions and studying advancements through all related impacts (both positive and negative). The experience of the ongoing COVID-19 pandemic will be written in future history books, and our actions and inactions will be judged, questioned, and assessed against what worked and what did not work. The future debates would entail a wide range of deficiencies, inequalities, and ineptitudes. Yet, we are uncertain if the future would be any better.

Central to our discussion is the term "urban health". We explore this in correlation with sustainability and peace, and we believe all three together shape a more holistic image of what the future should be. We start by questioning why urban health matters and delve into key ideological contestation over success and failures stories of our cities and communities. We question urban health because we witnessed many failures, inactions, incompetent behaviours, and irresponsible attitudes. At all levels, inactions were perceived. From global leadership and international cooperation to individual compliance with the simplest requirements, many failures show deficiencies in governance, institutional support, collective responses, etc. We faced that the rich got richer through many opportunities, and the poor became more vulnerable. The life values certainly differ between the two groups, as we witnessed piles of dead bodies on streets and rivers in poorer communities. We realised widespread neglect, disbelief, conspiracies, non-scientific information, and incapable minds. We witness how the rich accumulated more money and power in their never-ending race to reach the top and how governments' inactions led to distrusts and disbeliefs. We witnessed unhealthy ad unsustainable advancements, the absence of peace, and failures that led to long-term adversities. The COVID-19 will be remembered as a catastrophic time, as the day the world truly stopped.

In this chapter, we briefly explore urban health in relation to urban health equity and correlate it with global sustainability movements and strategies, like the ongoing Sustainable Development Goals (SDGs). These brief discussions will set a good foundation for the rest of the book, ensuring that we stress the importance of HSP nexus in cities and communities. Our goal is to put together the three overarching themes of "health", "sustainability", and "peace", with an intention to question our recent and current trends of development, show how unsustainable and unhealthy they are, and how peace is still far from our reach. We boldly question our imagination of a brighter future while not dealing with the challenges and problems of today. We see neglected opportunities at our doorsteps, knowing how important they are to make a tangible change for a healthier and more sustainable future. We discuss issues that have been diluted in major generic movements for a long, leading to empty promises, undefined goals, and counter-productive behaviours. The majority of us, as we know, are yet to be educated, and many governments and societies are yet to even reach the basic human needs. As the SDGs imply, the ultimate goal is to have a sustainable future for all, by achieving such an ideal future is not as easy as it sounds. We close this section with a remarkable poem by an anonymous social scientist (Quoted by Friedland, 1982; also see Cheshmehzangi, 2012), which reflects on what we believe is the situation now and ever:

> The reason why cities are ugly and sad,
> It is not that the people who live in them are bad
> It is just that the people who really decide,
> What goes on in the city live somewhere outside.

1.2 A Brief Discussion About "Urban Health Equity"

Urban health equity is an overarching topic, which related to multiple factors, such as urban environmental health hazards (Kjellstrom et al. 2007a), urban planning (Northridge and Freeman 2011), urban place (Coburn 2017), and urban food systems (Dixon et al. 2007. Urban health equity is developed based on multiple indicators, putting together integrated science, policy, and community needs (Corburn and Cohen 2012), which in fact is recognised as a major instrument for health policymakers. In more recent years, we see more discussion emerging in regards to climate change impacts on urban health, which include a wide range of public health implications (Fagliano and Diez Roux 2018). The implications for environmental health and equity are large, which also threatens global public health (Campbell-Lendrum and Corvalán 2007; Cheshmehzangi and Dawodu 2019). Hence, the suggestions are to promote health equity in order to enhance public health policy. There are also many social determinants in regards to health equity matters in urban areas, suggesting the benefits of urbanisation and urban development for health (Kjellstrom et al. 2007b; Kjellstrom and Mercado 2008; Barten et al. 2011). During the COVID-19 pandemic, urban health equity was discussed as part of the development of urban health policy, planning, and practice through an inclusive approach (Pineda and Corburn 2020). We know that for long urban health inequalities have become barriers to health equity, and hence we see a higher risk for exposure in poorer and vulnerable communities (LeBrón et al. 2019). Thus, there is an urgent need for action plans and better agenda (Friel et al. 2011), which would help to address multiple determinants of urban health equity in the long run.

Integrated approaches suggest a wide range of urban policy for health equity matters focused on using various data, such as environmental data, socio-demographic patterns, and geographical factors (Caiaffa et al. 2014). An example of such an approach is the urban health equity assessment and response tool, commonly known as Urban HEART, which was initially used by ten countries for pilot experimentation (Urban Heart 2010; WHO 2010). The tool reflects on major issues, such as inequalities in health among different communities and evaluating various determinants (Asadi et al. 2010; Najafi et al. 2013; Prasad et al. 2015). The tool is to aid "*decision-makers in planning action on inequities in health*" (Urban HEART 2010). Hence, there is room for action agenda development, leading to guidelines for urban planning and design solutions as well as pathways for health equity (Smit et al. 2011). The topic of urban health equity also brings together various stakeholders (Katz et al. 2015), including also "*hallmarks of a compromised process*" and "*towards the use of* ad hoc *participation processes*" (ibid). In healthy city planning, health equity has been promoted to improve urban social and environmental justice movements, policy development, experimentation, and towards the development of healthy cities (Corburn 2013). The same principles are also applied to promote health equity in poorer communities, such as slum upgrading projects (Corburn and Sverdlik 2017). This indicator-based approach has turned into an effective framework work-based approach, which is also suggested for spatial planning decision-making processes (Flacke and Köckler 2015), development of public health and healthcare infrastructures (Die Roux et al. 2020; Cheshmehzangi 2020b, 2021), and addressing inequality issues in more impoverished communities.

Although this book does not directly use "urban health equity" as a term, it aims to reflect on key indicators and determinants of achieving it in practice. Hence, it offers a multidimensional approach, which would indirectly relate to health equity matters, inequality issues, and deficiencies in cities and communities. Our arguments would be more than just a framework or tool and could lead to instrumental guidelines for future policies and practices. Some may lead to action plans, and some may influence multi-scalar agenda to achieve urban health for all. While urban health is not directly named as part of the SDGs, it is related to multiple dimensions of sustainable development. In the next section, we touch on some of these correlations.

1.3 Urban Health and Sustainable Development Goals (SDGs)

In their comprehensive analysis, Ramirez-Rubio et al. (2019) refer to urban health as an example of health in all policies, which relates directly to the implementation of SDGs. In their findings, they explicitly discuss the correlation of urban health to 38 targets across 15 SDGs. They also add four important aspects outside the SDGs, which include "*physical activity, noise, quality of life, and social capital*" (ibid). Their study also relates to an earlier report by the World Health Organization (WHO) (2016), which also addressed the importance of "*equitable healthier cities for sustainable development*". Their report also includes a range of discussions on correlations with SDG targets, including epidemic issues and reducing and preventing mortality and other disease infections.

By discussing health in the post-2015 era, Herrick (2016) explores two of the SDGs, which precisely focus on urban and health factors. In doing so, she argues further opportunities in these areas:

> ...a reinvigorated urban health is crucial not only to realize the urban and health SDGs, but also to capitalize on new, emergent research and opportunities that may emerge as global health work shifts to better reflect the contours of the SDGs.

In similar studies, we also see the role of socio-economic measures, which reflect on health equity and inequality issues. Examples such as metrics in urban health (Prasad et al. 2016; Saif-Ur-Rahman et al. 2018) use the SDGs for further debates on various health–sustainability perspectives, including also some of the SDG targets. In a more recent study, Azzopardi-Muscat et al. (2020) argue extensive relationships between urban design and health planning dimensions with SDG targets, also reflecting on some of the key challenges, practices, and policies that may need to go through revisions and potential public health reforms (Carmichael et al. 2019). Thus, we see tools like the urban HEART are getting closer to SDG dimensions (Prasad et al. 2018), which indicate a large range of determinants of health related to sustainable development strategies and practices.

While the correlation between (urban) health and sustainability is relatively strong in research and practice, there is a lack of discussions to include peace and prosperity in this domain. Only a handful of studies debate the topic in relation to relevant SDGs. An example is a study by Wesley et al. (2016), which argue health is not achievable without peace and explicitly suggests the importance of specific SDGs in achieving healthier communities. Other examples are related to WHO's health and peace initiative (WHO, 2020), which reflects on "*good health and well-being*", using the SDGs as their central point of discussion in achieving a healthy and peaceful society. Yet, some argue the challenges of implementing the SDGs and achieving peace, justice, and inclusive institutions (Hope 2020), by highlighting that sustained peace should be embedded in multiple SDGs. Such debates should hopefully generate more discussions on global political priority for urban health (Shawar and Crane 2017) and include more peace and prosperity indicators (Wamukoya et al. 2020). Therefore, in this book, we aim to bring together health, sustainability, and peace by exploring the context of cities and communities. The reason behind this exploration is mainly driven by the COVID-19 pandemic (and potential future pandemics) and its global impacts on societies, increasing inequalities, and many other challenges that help us to widen our horizon of urban health studies. Our aim is to propose pathways that include the healthiness of society, which also promotes sustainability and advocates peace and prosperity. If we achieve a model like this, we are on the right path to making a tangible change. It is then that our progress will be meaningful, and sustainability could mean a completely different thing.

1.4 Structure of the Book: Maintaining Health, Sustainability, and Peace in Cities

The book aims to put together debates on health, sustainability, and peace, which would help develop new ideas for a more comprehensive sustainable future. However, the paths we suggest are not for the future, but for today, for us to turn inactions to actions, deficiencies to efficiencies, counter-productivities to productivities, and unsustainable trends of development to sustainable and balanced models of development. The HSP nexus is important to achieve a healthier society, through which healthy cities and communities are slight possibilities. Our contribution to this direction is to support studies on urban health, urban sustainability, and peace (see Chap. 2). The book is structured in six parts to achieve these, each part dedicated to one of the six dimensions of physical, environmental, economic, social, institutional, and technical. Before going to dimensions, Chap. 2 of the book focuses on the correlations between health, sustainability, and peace, where we precisely discuss the topic of positive peace and its primary pillars. Afterward, Chaps. 3–20 will serve as an exploration of 18 aspects under the six mentioned dimensions. Thus, in each dimension, we highlight three primary aspects. These selected aspects are related to sustainable development discourse, healthy cities, and primary pillars of positive peace. Some are directly related to urban sustainability and health matters, and some are directly part of positive peace pillars. We aim to find crossovers between the three themes and debate the HSP nexus in all of them. Finally, each aspect is structured in four sections, first through some initial general viewpoints, second on how cities reacted during the pandemic, third through a reflective discussion in responding to issues and challenges of each aspect, and fourth a brief summary. Altogether, 18 aspects are discussed in this book, which stipulates a comprehensive understanding of the HSP nexus for cities and communities. The book finally concludes with further debates on "the city we need" and expands on viewpoints on the HSP nexus. In doing so, we also reflect on some of the issues, challenges, lessons, and future research pathways (see Chap. 21). Some recommendations are also made to ensure our contribution is tangible and valuable for future directions in scholarly research, policy development, and practice.

Here, we summarise all six parts of the book.

Part I—Under the physical dimension, three selected primary aspects are as follows:

- Urban density and land use planning (Chap. 3);
- Resilient urban form (Chap. 4);
- Critical infrastructures (Chap. 5).

Part II – Under the environmental dimension, three selected primary aspects are as follows:

- Well-functioning government and peaceful living environments (Chap. 6);
- Essential systems for primary city operations (Chap. 7);
- Resource efficiency and distribution (Chap. 8).

Part III—Under the economic dimension, three selected primary aspects are as follows:

- Healthy business environment (Chap. 9);
- Low level of corruption (Chap. 10);
- Supply chain resilience (Chap. 11).

Part IV—Under the social dimension, three selected primary aspects are as follows:

- Information sharing for a healthier society (Chap. 12);
- Human capital and equity (Chap. 13);
- Social and public life through a public realm perspective (Chap. 14).

Part V—Under the institutional dimension, three selected primary aspects are as follows:

- Public rights for social and public services (Chap. 15);
- Good relations with neighbours (Chap. 16);
- Public networks and governance matters (Chap. 17).

Part VI—Under the technical dimension, three selected primary aspects are as follows:

- Technology availability and accessibility (Chap. 18);
- Asset and facilities management (Chap. 19);
- Smart platforms and technical solutions for smart-resilient models (Chap. 20).

Finally, we hope the book is a starting point for future research related to the HSP nexus in cities and communities. We trust the COVID-19 pandemic could open up new research opportunities to guarantee the development of game-changing policies, global agenda, sustainable pathways, and new paradigms. We believe innovations are yet to blossom, and fruitful interventions are yet to emerge as novel best practices. We hope we could learn a lot from the day the world stopped.

References

Asadi LM, Vaez MMR, Faghihzadeh S, Montazeri A, Kalantari N, Maher A et al.(2010). The application of urban health equity assessment and response tool (Urban HEART) in Tehran; concepts and framework. Available from: https://www.sid.ir/en/journal/ViewPaper.aspx?id=204145

Azzopardi-Muscat N, Brambilla A, Caracci F, Capolongo S (2020) Synergies in design and health. The role of architects and urban health planners in tackling key contemporary public health challenges. Acta Biomed 91(3-S):9–20

Barten F, Akerman M, Becker D, Friel S, Hancock T, Mwatsama M et al (2011) Rights, knowledge, and governance for improved health equity in urban settings. J Urban Health 88(5):896–905

Caiaffa WT, Friche AAL, Dias MAS, Meireles AL, Ignacio CF, Prasad A, Kano M (2014) Developing a conceptual framework of urban health observatories toward integrating research and evidence into urban policy for health and health equity. J Urban Health 91 (1):1–16

Campbell-Lendrum D, Corvalán C (2007) Climate change and developing-country cities: implications for environmental health and equity. J Urban Health 84 (1):109–117

Carmichael L, Townshend TG, Fischer TB, Lock K, Petrokofsky C, Sheppard A, Sweeting D, Ogilvie F (2019) Urban planning as an enabler of urban health: challenges and good practice in England following the 2012 planning and public health reforms. Land Use Policy 84:154–162

Cheshmehzangi A (2020) The city in need: urban resilience and city management during disruptive disease outbreak events. Springer, Singapore

Cheshmehzangi A (2020) Reflection on early lessons for urban resilience and public health enhancement during the COVID-19. Health 12(10):1390

Cheshmehzangi A, Dawodu A (2019) Sustainable urban development in the age of climate change—people: the cure or curse. Palgrave Macmillan, Singapore

Cheshmehzangi A (2012) Reviving the city and identifying approaches to sustain growth. In: Proceedings of an international conference on masterplanning the future: modernism: east, west, & across the World, held in Suzhou, China, October 2012, pp 31–44

Cheshmehzangi A (2021) Vulnerability of the UK's BAME communities during COVID-19: the review of public health and socio-economic inequalities. J Hum Behav Soc Environ, pp 1–18

Corburn J (2013) Healthy city planning: from neighbourhood to national health equity. Routledge, New York and London

Corburn J (2017) Urban place and health equity: critical issues and practices. Int J Environ Res Public Health 14(2):117

Corburn J, Sverdlik A (2017) Slum upgrading and health equity. Int J Environ Res Public Health 14(4):342

Corburn J, Cohen AK (2012) Why we need urban health equity indicators: integrating science, policy, and community. PLoS Med 9(8):e1001285

Dixon J, Omwega AM, Friel S, Burns C, Donati K, Carlisle R (2007) The health equity dimensions of urban food systems. J Urban Health 84(1):118–129

Fagliano JA, Diez Roux AV (2018) Climate change, urban health, and the promotion of health equity. PLoS Med 15(7):e1002621

Flacke J, Köckler H (2015) Spatial urban health equity indicators–a framework-based approach supporting spatial decision making. Sustain Dev Plan VII 193:365–376

Friedland R (1982) Power and crisis in the city: corporations. Macmillan Press Ltd., Unions and Urban Policy, London

Friel S, Akerman M, Hancock T, Kumaresan J, Marmot M, Melin T, Vlahov D (2011) Addressing the social and environmental determinants of urban health equity: evidence for action and a research agenda. J Urban Health 88(5):860–874

Herrick C (2016) An urban health worthy of the post-2015 era. Environ Urban 28(1):139–144

Hope KR Sr (2020) Peace, justice and inclusive institutions: overcoming challenges to the implementation of sustainable development goal 16. Glob Change Peace Secur 32(1):57–77

Katz AS, Cheff RM, O'Campo P (2015) Bringing stakeholders together for urban health equity: hallmarks of a compromised process. Int J Equity Health 14(1):1–9

Kjellstrom T, Mercado S (2008) Towards action on social determinants for health equity in urban settings. Environ Urban 20(2):551–574

Kjellstrom T, Mercado S, Sami M, Havemann K, Iwao S (2007) Achieving health equity in urban settings. J Urban Health 84(1):1–6

Kjellstrom T, Friel S, Dixon J, Corvalan C, Rehfuess E, Campbell-Lendrum D et al (2007a) Urban environmental health hazards and health equity. J Urban Health 84(1):86–97

LeBrón AM, Torres IR, Valencia E, Dominguez ML, Garcia-Sanchez DG, Logue MD, Wu J (2019) The state of public health lead policies: Implications for urban health inequities and recommendations for health equity. Int J Environ Res Public Health 16 (6):1064

Najafi B, Arzaghi M, Fakhrzadeh H, Sharifi F, Shoaei S, Alizadeh M et al (2013) Mental health status and related factors in aged population: Urban health equity assessment and response tool (Urban-HEART) study in Tehran. Iran J Diab Metab 13(1):62–73

Northridge ME, Freeman L (2011) Urban planning and health equity. J Urban Health 88(3):582–597

Pineda VS, Corburn J (2020) Disability, urban health equity, and the coronavirus pandemic: promoting cities for all. J Urban Health 97(3):336–341

Prasad A, Kano M, Dagg KAM, Mori H, Senkoro HH, Ardakani MA et al (2015) Prioritizing action on health inequities in cities: An evaluation of Urban Health Equity Assessment and Response Tool (Urban HEART) in 15 cities from Asia and Africa. Soc Sci Med 145:237–242

Prasad A, Gray CB, Ross A, Kano M (2016) Metrics in urban health: current developments and future prospects. Annu Rev Public Health 37:113–133

Prasad A, Borrell C, Mehdipanah R, Chatterji S (2018). Tackling health inequalities using urban HEART in the sustainable development goals era

Ramirez-Rubio O, Daher C, Fanjul G, Gascon M, Mueller N, Pajín L et al (2019) Urban health: an example of a "health in all policies" approach in the context of SDGs implementation. Glob Health 15 (1):1–21

Saif-Ur-Rahman KM, Anwar I, Hasan M, Hossain S, Shafique S, Haseen F, Khalequzzaman M, Rahman A, Islam S (2018) Use of indices to measure socio-economic status (SES) in South-Asian urban health studies: a scoping review. Syst Rev 7(1):1–9

Shawar YR, Crane LG (2017) Generating global political priority for urban health: the role of the urban health epistemic community. Health Policy Plan 32(8):1161–1173

Smit W, Hancock T, Kumaresen J, Santos-Burgoa C, Meneses RSK, Friel S (2011) Toward a research and action agenda on urban planning/design and health equity in cities in low and middle-income countries. J Urban Health 88(5):875

Urban HEART (2010) Urban health equity assessment and response tool user manual. Available from: http://citeseerx.ist.psu.edu/viewdoc/summary?doi=10.1.1.365.6512

Wamukoya M, Kadengye DT, Iddi S, Chikozho C (2020) The Nairobi urban health and demographic surveillance of slum dwellers, 2002–2019: value, processes, and challenges. Glob Epidemiol 2:100024

Wesley H, Tittle V, Seita A (2016) No health without peace: why SDG 16 is essential for health. The Lancet 388(10058):2352–2353

WHO (World Health Organization) (2010) Urban HEART: urban health equity assessment and response tool. WHO Centre for Health Development, Kobe, Japan, 48 pages, Available from: https://apps.who.int/iris/bitstream/handle/10665/79060/9789241500142_eng.pdf

WHO (World Health Organization) (2016) Global report on urban health: equitable healthier cities for sustainable development. 239 pages, Available from: https://apps.who.int/iris/handle/10665/204715

WHO (World Health Organization) (2020) Health and peace initiative. 48 pages, Available from: https://apps.who.int/iris/bitstream/handle/10665/332938/9789240005792-eng.pdf

2 Signifying the Correlation Between Urban Health, Sustainability, and Positive Peace

Nothing really changes; it only evolves for better or worse.
—Ali Cheshmehzangi (2017)

2.1 The Significance of Inequalities and Facing the Pandemic

During the ongoing COVID-19 pandemic, we noticed many differences between cities and communities facing and dealing with the adversities of the event. While the author believes we do not all experience the same situation, it is important to highlight a range of actions and inactions that led to the success and failures of managing the outbreak at the city and/or community levels. We note that the situation was not experienced the same across the board. In fact, the pandemic highlighted significant disparities between countries/contexts of different conditions, cities of different status, communities of different socio-economic demographics, ethnicities, and different jobs and incomes, genders of different daily needs, and various age group factors. In fact, we clearly see that there are inequalities across the board and facing the situation depends on multiple variables. The inequalities were perceived since the inception and all the way through to vaccine acquisition, distribution, and treatments. As the book is completing, the situation persists, reminding us of the wide range and wide-scale inequalities that remain to be resolved.

Metaphorically, if we consider the pandemic as a storm, we see that everyone experienced the same storm differently. While the storm crushed the economies of poorer communities, we noticed that some of the rich became richer during the pandemic. We also noticed the inequalities that exist between different groups of people, ethnics, and backgrounds (Cheshmehzangi 2020a), which has been the major highlight of this pandemic that is rooted for a long in our societies (Bambra et al. 2020; Bowleg 2020; Engzell et al. 2020; Dang and Nguyen 2021; Fisher and Ryan 2021; Su et al. 2021). Thus, the exacerbating inequalities continue to become the major longer-term issues of the pandemic (Cheshmehzangi 2020b; Van Dorn et al. 2020). There is also enough evidence that shows inequality could spread the pandemic (Ahmed et al. 2020) or indicate that the vulnerable groups are often forgotten (Patel et al. 2020). More importantly, we see differences that vary differently between locations for different groups of people, showing the impact of structured and longer-term inequalities in facing the pandemic. One example is the ongoing health inequalities (Abedi et al. 2020), which suggests the complex situation of social capital failures and income inequalities (Elgar et al. 2020). The existing literature also highlights other inequalities, such as

A. Cheshmehzangi, *Urban Health, Sustainability, and Peace in the Day the World Stopped*, Sustainable Development Goals Series, https://doi.org/10.1007/978-981-16-4888-5_2

racial inequalities (Kim and Bostwick 2020), family and employment issues (Blundell et al. 2020), response inequalities (Okoi and Bwawa 2020), and all the way to digital inequalities (Beaunoyer et al. 2020) that exist in our contemporary society.

So far, one of the key aspects that were largely debated was the lack of equity between different groups of communities. As a result, some were more capable of managing the pandemic, and some were not. This applies to multiple spatial levels of countries, provinces/states/counties, regions, cities, and communities. The end results vary very differently, with some facing longer-time adversities, starting from longer (and multiple) lockdowns, to longer processes of recovery, treatment, vaccine distribution, risk management, etc. However, the situation persists as the longer economic impacts are yet to be assessed and recovered (Cheshmehzangi 2020b), meaning that we are likely to face much longer hardships.

2.1.1 Beyond the Three Primary P's: Planning, Policies, and People

Here, we highlight three primary P's, namely "planning", "policies", and "people"—in this orderly manner. The planning aspect comes from the perspective of preparedness (Cheshmehzangi 2020b), reflecting on the health of cities and communities. Cities and countries that managed the outbreak more successfully are often the ones known to have more robust mechanisms for better planning. This could be related to their planned infrastructures, planned services, planned mechanisms, etc. While preparedness is part of planning, we noted that some countries or cities fail to follow their original preparations that were meant to boost their responsiveness to the pandemic. In such cases, the expected interventions were minimised or became failures. Therefore, policies played a significant part in sustaining robust responsiveness throughout the pandemic. Like any other outbreak situation, policies are essential to help set up protocols, processes, and a clear response plan (WHO 2020; Arain et al. 2021). A systematic approach has proven to be a successful model for better preparedness and transition strategy (Biswas et al. 2020), implementing global guidance for specific sectors (Collaborative 2020) and the value of policies (Rodriguez-Morales et al. 2020). Some of these policies could be regarded as unprecedented policies, which are often designed for specific reasons (Hsiang et al. 2020). Some policies, on the other hand, focus on key aspects of suggesting guidelines for preparedness (Mossa-Basha et al. 2020), specific measures and activities (Dell'Ariccia et al. 2020; Piguillem and Shi 2020), as well as a range of mandatory and voluntary policies (Betsch et al. 2020). In all cases, we notice the effectiveness of various measures and policies for enhancing society's safety and controlling the pandemic. Therefore, policies play a major part in containing the outbreak and reopening businesses and specific sectors (Desai and Aronoff 2020; Ho et al. 2020). In most successful containment or rapid management examples, we see that the availability of policies is not the only measure. Therefore, we need to ensure policies are successfully developed, implemented, and followed by the relative stakeholders, including people. This leads us to the last primary "P", meaning people's compliance and support to respond to planning and policies set for managing the outbreak. If all three are in place, then the success rate of managing the outbreak/pandemic is expected to be high. Nonetheless, the situation is more complicated than just having them in place. The situation is often faced with other aspects that reflect on the healthiness of cities and communities.

Therefore, in this chapter, we go beyond these three primary P's and focus more on what becomes the novelty of the book, i.e. discussing the correlation between urban heath, sustainability, and positive peace. At first, we briefly introduce the concept of positive peace, commonly known as peace with justice for all. This introduction highlights eight dimensions of positive peace and how they relate to healthy cities or "urban health", as the book implies. Afterwards, we introduce six sustainability dimensions for healthy cities and discuss the correlations that exist between urban health, sustainability, and positive peace. The chapter then concludes with a general overview of what becomes the basis of the following six parts in the following 18 chapters.

2.2 Introduction to Positive Peace for Healthy Cities

According to Grewal (2003), the word peace is misused (and even abused) in the literature, mainly because it is partially unreal. There are also complications in conceptualising the term and agreeing on what it means in different disciplines and contexts. Hence, we see the definition of complexity is partly resolved by distinguishing between negative peace and positive peace. In this regard, we summarise what Barash (2017) suggests in terms of defining the two (apart from cultural peace), which is also similar to other definitions by other scholars, such as Reardon (1988), Roberts (2008), Clark (2009), Galtung (2011), and Galtung and Fischer (2013):

> The pursuit of peace seeks not just to achieve the "negative peace" of halting warfare and direct violence, but also to nurture the "positive peace," of life-affirming policies and conditions that promote harmony and social justice (Barash 2017).

Coined by Galtung (1964), the term positive peace relates to "*the presence and prevalence of positive attributes, conditions, and priorities*" that should ideally promote "*social and economic justice, environmental integrity, human rights, and development*" and "*contribute to the structural integration of human society*" (also see Amadei 2021). This complex definition is completed by Fischer's (2007) view of positive peace activities, which include "*building a life-sustaining economy at the local, national, and global level in which everyone's basic needs are met*", "*good governance and participation*", "*self-determination*", and "*human rights*". These indeed directly link with the urban health ideals, which suggest the health and well-being of society as the priority.

Moreover, Shields (2017) suggests the limits of negative peace, hence arguing in favour of the state of positive peace. For the health of cities and communities, in particular, we see the role of positive peace as more effective as it represents opportunities for stability or sustainability and achieving transitional justice in society (Sharp 2011). Set aside the obvious differences between various violent situations (Grewal 2003), we refer to positive peace as the backbone of achieving better policies for better inequalities and social justice. As a building framework for positive peace, the existing literature highlights key factors of social and economic rights, peacebuilding agenda, and achieving human security in society (Cahill-Ripley 2016). More importantly, as said earlier, we see a close link between the healthiness of cities and communities and sustaining positive peace in society.

2.2.1 Positive Peace Pillars or Domains

The Institute for Economic Peace (IEP), located in Sydney, Australia, proposed the idea of the Positive Peace Index (PPI), referring to positive peace as "*the attitudes, institutions, and structures that create and sustain peaceful societies*" (IEP 2017; Amadei 2021). Also earlier, Amadei (2020) suggests the idea of the peace triangle (developed by Galtung 1990), including cultural peace, negative peace, and positive peace, could prove that the three overlap, "*representing the extent of the enabling environment in which peace unfolds over time*" (Amadei 2020). Therefore, we could see that the IEP's approach to positive peace is, in fact, to distinguish between eight interdependent eight pillars or

Fig. 2.1 Demonstration of eight pillars or domains of Positive Peace Index (2017, 2018 *Source* Drawn by the Authors, according to IEP)

domains, which provide a comprehensive image of the optimum positive peace environment (IEP 2017, 2018; Amadei 2020).

According to IEP (2017, 2018), the eight pillars/domains are defined as (1) well-functioning government, (2) equitable distribution of resources, (3) free flow of information, (4) good relations with neighbours, (5) high level of human capital, (6) acceptance of the rights of others, (7) low levels of corruption, and (8) sound business environment (Fig. 2.1). Also developed by IEP (2017), the following indicators are verified for each pillar:

(1) **Well-functioning Government (WFG)**
Primary Indicators: Democratic political culture, government effectiveness, and rule of law;
(2) **Equitable Distribution of Resources (EDR)**
Primary Indicators: Inequality-adjusted life expectancy, poverty gap, and social mobility;

(3) **Free Flow of Information (FFI)**
Primary Indicators: Freedom of the Press Index overall score, mobile phone subscription rate, and World Press Freedom Index overall score;
(4) **Good Relations with Neighbours (GRN)**
Primary Indicators: Hostility to foreigners, number of visitors, and regional integration;
(5) **High Level of Human Capital (HLC)**
Primary Indicators: Secondary school enrolment, Global Innovation Index, and Youth Development Index overall score;
(6) **Acceptance of the Rights of Others (ARO)**
Primary Indicators: Empowerment index, group grievance rating, and gender inequality;
(7) **Low Levels of Corruption (LLC)**
Primary Indicators: Factionalised elites, perceptions of corruption score, and control of corruption.
(8) **Sound Business Environment (SBE)**
Primary Indicators: Business environment, economic freedom overall score, and GDP per capita.

In this book, it is important for us to understand the positive peace pillars as part of achieving urban health and sustaining it. Therefore, what we discuss throughout the next 18 chapters will include all the pillars but also other aspects for each sustainability dimension, which would ultimately help us achieve health and sustainability of cities and communities. In this regard, we clearly argue that positive peace is just part of a bigger picture, ensuring that health and well-being are maintained and sustained, especially during and after the case of pandemics. This is the reason why we include the overarching term "peace", rather than just "positive peace". For us, the above indicators are also expanded or interpreted in how cities develop and operate, ensuring that all pillars are remained and valued as part of maintaining or sustaining the overall health–sustainability–peace nexus in cities around the world.

A good example is a study that looks into the already established positive peace and sustainability nexus, which is suggested to go beyond the triple bottom line of sustainability (Bond 2014). Also, the current debates on sustainability–peace nexus in the context of climate change (Virji et al. 2019) argue the role of social harmony and justice in achieving the SDGs. They also suggest the significance of this nexus and opportunities for potential strategies and policies to promote peace and security for sustainability (see Brauch et al. 2016). This topic is further assessed through a scientometric analysis (Sharifi et al. 2021), through which we can see a co-occurrence network of two keywords of "sustainability" and "peace", which again is rooted in some of the earlier work by IEP (2019a; b) and some other institutes and organisations that explore the nexus in a more effective way. In addition, Sharifi et al. (2021) put together solid evidence that shows significant interaction between the pillars of positive peace and the SDGs. Again, their investigation investigates the linkages and connections across different disciplines, allowing for more opportunity for interdisciplinary research.

2.2.2 Positive Peace Dimensions and Sustainability

In their studies, Milesi and Franco (2020) develop a method to building more peaceful, inclusive, just, and accountable societies. In their work, they propose four steps of observe, plan, act, and reflect, which are also correlated with key Sustainable Development Goals (SDGs). They specifically relate positive peace to SDG #16, which is defined as "Peace and Justice". They then consider the inclusion of at least seven on the SDGs, including "No Poverty" (SDG #1), "Quality Education" (SDG #4), "Gender Equality" (SDG #5), "Good Jobs and Economic Growth" (SDG #8), "Reduced Inequalities" (SDG #10), "Sustainable Cities and Communities"

(SDG #11), and "Partnerships for the Goals" (SDG #17). By connecting the positive peace dimensions of change to SDG #16, they explicitly study four levels of no direct violence, structural justice (related to governance), structural justice (related to poverty and inequality), and cultural understanding. In each of these, specific targets and indicators are also suggested, which also directly relate to the seven relevant SDGs.

Recent scholarly work also includes the consideration of integrated sustainability to develop a theoretical framework that connects the environment, peace, and sustainability (Fisher et al. 2021). This approach indicates the growing importance of the sustainability agenda and towards the framing of sustainable development (Kidd 1992; Clune et al. 2020; Fisher et al. 2021). In formalising a framework for integrated sustainability, Fisher et al. (2021) also suggest three propositions, which include basic human needs, subjective values and equity, and safeguarding a variety of environmental, social, and public goods within our social–ecological systems. These are then followed by the fourth proposition, which suggests the following:

> Sustainable development involves the process of resolving the inherent incompatibilities between human development and ecological integrity through institutions that facilitate cooperation and regulate competition in social-ecological systems (ibid).

In this example, we note multiple variables that should be taken into consideration for each of the propositions. In doing so, we are then able to see a model that considers two factors of coordinated cooperation and regulated political competition (ibid), both closely related to the overarching nexus between environment, peace, and sustainability. Furthermore, in regard to indirect forms of violence, regarded as positive peace, we can consider methods of achieving societal well-being (Galtung 1964; Fisher et al. 2021), which again are important for balancing better social harmony and cooperation (Galtung 1969). In their arguments, Fisher et al. (2021) suggest that integrating the positive dimensions of peace with sustainability enables us to promote social values and aspects such as "*individual agency, equity, and opportunity for human flourishing*". In this regard, we can confirm the correlation between positive peace and sustainability is robust and includes key cooperation, promotions, competitions, regulations, mechanisms, as well as well-being goals. In their framework (ibid), these aspects are included to ensure an integrated sustainability model is attained.

Another example is a recent work by Amadei (2021), which again systematically studies the nexus between sustainability and peace. He explicitly explores the differences between positive and negative peace (also see Galtung 1964; Fischer 2007) before delving into the PPI discussions and its dimensions proposed by the IEP. Following the studies of Sharifi et al. (2021), we can corroborate the substantial existing sustainability–peace nexus in the field. Still, we also aim to expand to connect this nexus with the topic of urban health or healthy cities. In doing so, we introduce sustainability dimensions for healthy cities before closing the chapter with a further opinion on this triangle correlation, i.e. health–sustainability–peace.

2.3 Introduction to Sustainability Dimensions for Healthy Cities

In general, a healthy city should include healthy people, healthy environments, and healthy communities (Planning Tank 2017). Nonetheless, the sustainability of health initiatives is still not yet matured enough, especially in urban studies and urbanism. This is why we aim to study the triangle correlation that suggests the importance of urban health by achieving and sustaining well-being in society. Amadei (2021) mentioned that peace appears in several SDGs, and health is one of those that is also related to other SDGs (see UN 2019). Amadei (2021) also debates the relevance of "*human development concepts of peace and sustainability*" to critical concepts of health and resilience, which are generally hard to conceptualise. It is essential to note the two terms of "health" and "resilience" often come together, particularly that the two are interrelated when we study healthy cities and/or urban health. There is certainly no fits-all united model that could define an ideal healthy city, but we rest assured that the sustainability–peace nexus could play a major part in maintaining and sustaining social health, societal health, and urban health.

Central to this book, the ongoing COVID-19 pandemic has given a good reason to study urban health in more detail. To do so, we may need to look into transitions (Cheshmehzangi 2021a) and potential paradigm shifts through revisiting the built environments (Cheshmehzangi 2021b) and cities and communities around the world. The study of healthy cities on their own may include six areas of governance, sustainability, determinants (of health), strategies, equity, and evidence (de Leeuw and Green 2017). But, the study of the health–sustainability–peace (HSP) nexus could help us widen our perspectives regarding achieving better resilience and well-being, as well as a multidisciplinary view of city operations, urban systems, and urban sustainability. In this regard, we explore the health–sustainability–peace nexus from six sustainability dimensions, including (1) physical, (2) environmental, (3) economic, (4) social, (5) institutional, and (6) technical (Fig. 2.2). We hope that by following a more comprehensive approach, we could suggest more aspects and pathways towards healthier and more resilient cities and communities. In doing so, we realise the importance of not only the usual three or four dimensions of sustainability but also the inclusion of physical and technical aspects that are increasingly important in developing sustainable and healthy cities.

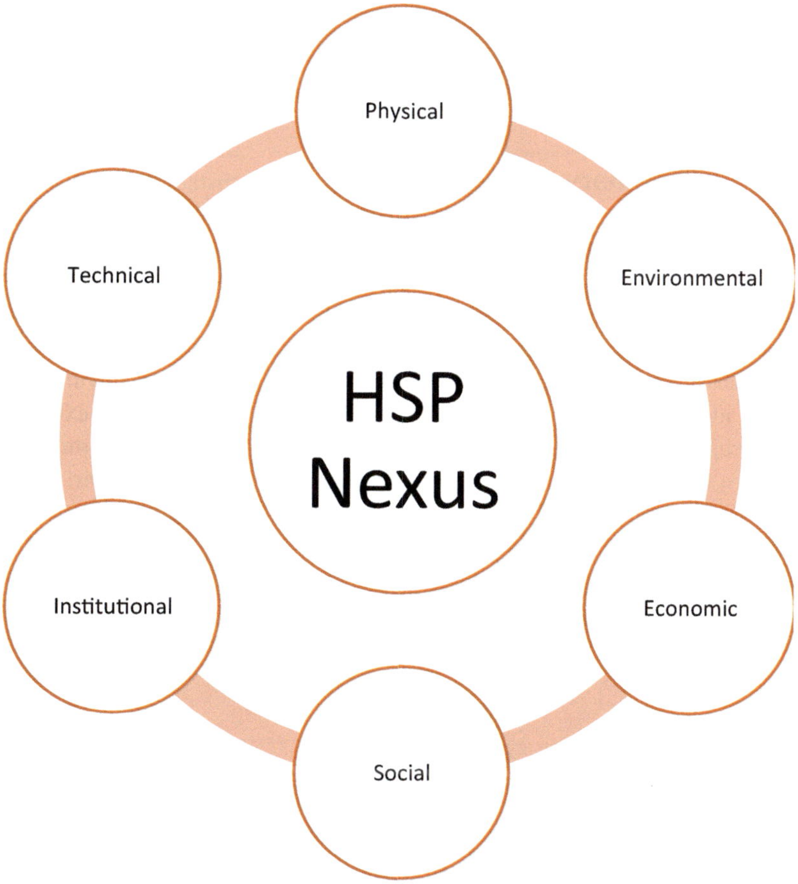

Fig. 2.2 Selected six dimensions as part of the health–sustainability–peace (HSP) nexus expanded from the original 3–4 dimensions of sustainability and towards a comprehensive understating of healthy cities

In this book, the HSP nexus is important to better evaluate and reflect on the current situations of cities and communities in facing the pandemic. This enables us to better investigate the resilience of cities, consider and avoid particular development trends, scrutinise the unhealthy situations in cities, find alternative ways or indicators for better urban health, and identify the role of sustainability–peace factors in achieving the ultimate urban health. As shown in Fig. 2.2, the six proposed dimensions are expected to detect health problems in cities, enabling us to be more reflective after the experience of the COVID-19 pandemic. As mentioned earlier in the introduction of positive peace pillars, we also see several dimensions that are summarised as various conditions, such as "systematic and complex", "virtuous or vicious", "preventive", "underpinning resilience and non-violence", "informal and formal", and "supporting development goals" (Positive Peace, n.d.). All six dimensions relate to our proposed sustainability dimensions, indicating very carefully the consideration of complex urban situations, planning, adaptive thinking, and addressing the SDGs.

2.4 A Summary: Correlation Between Health, Sustainability, and Peace (HSP)

This chapter highlights the potential correlation that exists between urban health, sustainability, and peace (mainly positive peace here). What we have identified so far is the strong relationship between sustainability and peace, which has proven to be a growing area of research in recent years. The linkages of this nexus with urban health enable us to further extend the discussions on the resilience and well-being of cities and communities. In doing so, we prove the HSP nexus is suitable for further consideration, especially if we aim to reflect on the impacts of COVID-19 on cities and communities. As a major direction, the ultimate goals are to revisit the current development trends, evaluate what worked and what did not work during the pandemic, and target to enhance the resilience of cities. Hence, the HSP nexus is inevitable as urban health is the precursor to healthy people and healthy communities.

This chapter serves as a brief introduction to the following six parts of the book, which are focused on the mentioned six dimensions (see Sect. 2.3). First, we will start with the physical dimension, which includes key areas that help to enhance the healthiness and well-being of communities. Afterwards, we delve into the triple bottom line of sustainability, namely environmental, economic, and social dimensions. Then, we further expand the HSP nexus discussions by exploring key areas of the institutional dimension. Lastly, we highlight the emerging "technical dimension", which has proven to become an important dimension in achieving urban health, sustainability, and peace. For all dimensions, we highlight three specific areas that relate to our proposed HSP nexus. These are studied individually in facing the pandemics or, in this case, "the day the world stopped".

Before we move into the main body of the book, we provide several cases of concern related to the HSP nexus and the COVID-19 impacts. These are summarised in Boxes 2.1 to 2.20, before the start of our discussions in the following six parts.

BOX 2.1 Corruption and COVID-19

The issues of corruption during the COVID-19 pandemic are widespread. Thus, we see more threats from corrupted conditions than the actual pandemic itself. From media dissemination to politicisation of the pandemic, we see that COVID-19 has created many opportunities for corruption. Not only that it truly threatens democracies, but also it has sabotaged peace in many countries. The lack or absence of robust institutional structures often leads to such situations. This pandemic showed two sides of how things happened on the ground. On one side, we see fundamental rights and freedom viewpoints conflicting with decision-making processes and the containment process of the disease. On the other side, we see the rise of authorisation, disputing the public rights, and yet make substantial progress in containing the outbreak. We see actions and inactions in both, and we witnessed many examples of corruption at work, management, governance, public services, social services, etc. Finally, we note that a global movement is needed to respond to corrupted cycles of growth and development, enhance transparency, and strengthen anticorruption efforts. We see larger-scale corruption issues that are not necessarily surprising from the allocation of designated response funds to their distribution and fair use. The world needs to wake up and restructure its institutions, institutional support mechanisms and channels, and relationships between them. Or else, the United Nations should not be concerned about corruption in a time of crisis. But as shown below, this is a big concern at all levels.

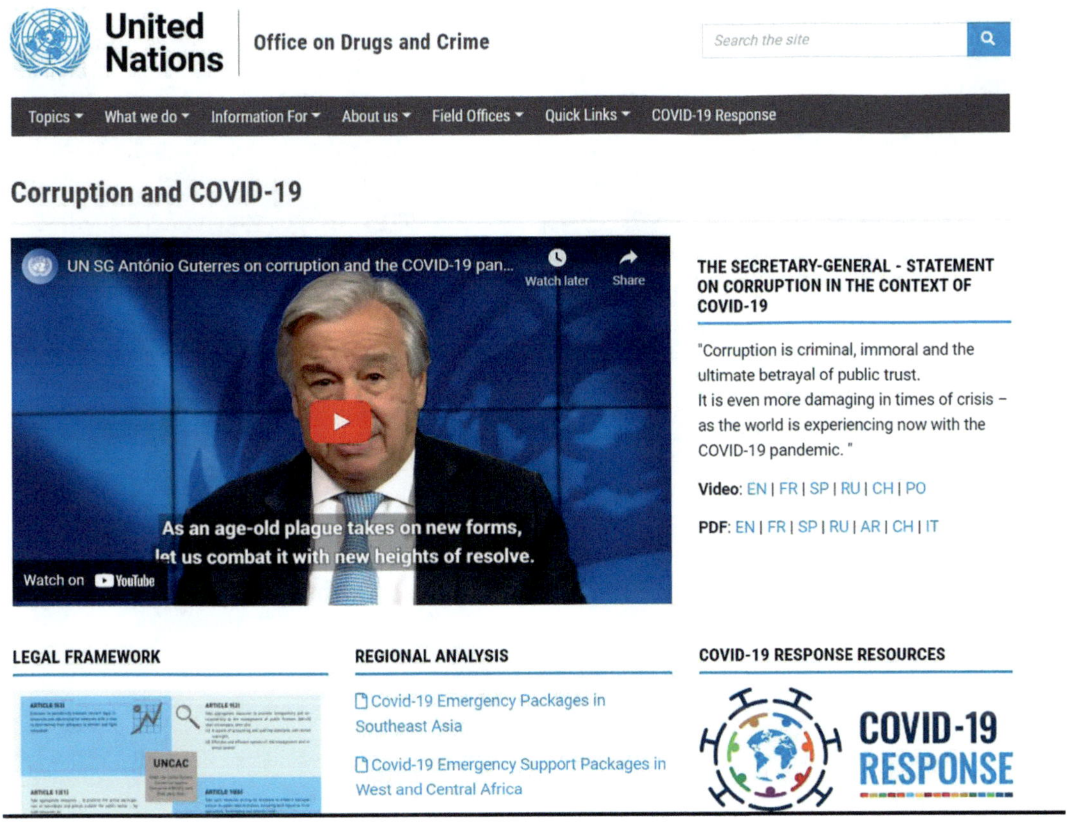

BOX 2.2 COVID-19 and Peace

Many reports were developed to ensure peace is maintained during the pandemic crises. Some examples are shown here, indicating the central role of peacebuilding practices and their importance in dealing with political situations, socio-economic issues, and related conflicts. From IEP to the UN departments, these examples provide specific guidelines to combat conflict issues and maintain peace.

BOX 2.3 COVID-19 and Addressing Inequalities

As part of peacebuilding practices, inequality issues were carefully highlighted. These examples below indicate various guidance notes and documents/reports, specifically related to particular societal inequalities. From gender inequality issues to the development of security, we see documents developed that respond to context-specific problems and some that focus on general investment and transition processes. In these cases, as shown below, we see examples of documents and guidance reports that are led by the United Nations agencies, collaborations between industry and education, non-governmental organisations (NGOs), policymakers, businesses, etc.

BOX 2.4 Global Peace Index (GPI) and Its Relevance to the COVID-19 Pandemic

The results this year show that the level of global peacefulness deteriorated, with the average country score falling by 0.34%. This is the ninth deterioration in peacefulness in the last twelve years, with 81 countries improving, and 80 recording deteriorations over the past year. The 2020 GPI reveals a world in which the conflicts and crises that emerged in the past decade have begun to abate, only to be replaced with a new wave of tension and uncertainty as a result of the COVID-19 pandemic".

Source: Global Peace Index 2020, Provided by "reliefweb", Extracted from: https://reliefweb.int/report/world/global-peace-index-2020.

BOX 2.5: COVID-19 Impact on Vulnerable Groups and Communities

Reflecting on vulnerabilities was among the first things in dealing with peace-building processes in cities and communities. One of the main focuses has been on poverty issues, growing marginalisation, and increasing vulnerability. The many reports and guidance documents, as shown below, are representative of inequality issues that suggest widening vulnerabilities in certain groups. Since the inception of the COVID-19 pandemic, we see growing problems related to equity and equality matters. The (pretending) claim that all of us are in the same boat is entirely false and only delusional for those trying to neglect the widening issues of vulnerable groups and communities. We are in no way in the same boat, and as said in this book, we are not even experiencing the same storm. The below examples are just a few reports that explore vulnerability in specific groups, nations, and communities.

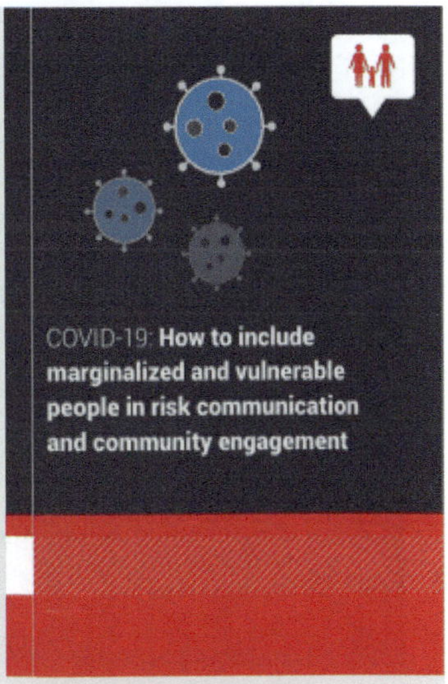

BOX 2.6: COVID-19 and Related Protocols

The COVID-19 pandemic has enabled us to develop specific guidelines regarding dealing with such health crises, associated protocols, and procedures that were not in place before. Many of these depend on particular contexts and include a good variety of examples that other nations could replicate or adapt. Many of these guidelines and protocols were made openly available to the public or are on websites and/or are advertised widely in streets, workplaces, public places, etc. The below examples are just a few for us to understand context-specific platforms and generic protocols that could help poorer countries or regions, as well as helping to contain the disease and provide support to people, businesses, organisations, and working environments.

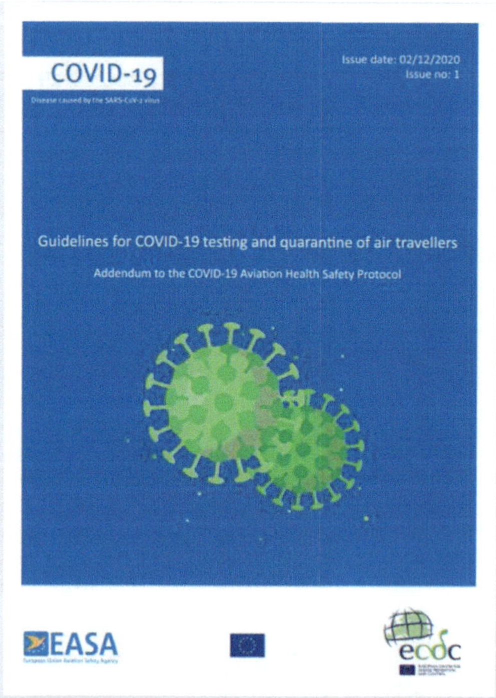

BOX 2.7 Education Matters: Issues, Risks, and Impacts

Education and educating have been the backbones of dealing with the pandemic. Lack of education is commonly seen across the world, undermining the progress of achieving the SDG targets. A significant issue during the COVID-19 pandemic has been the growing digital divide in many countries. The segregations, lack of education, and the ever-increasing divide continuously affect the more impoverished communities, vulnerable groups, and those with digital knowledge. From policy reports to guideline documents, the issues of the digital divide have been highlighted widely. There have been many webinars, guidelines, and sharing sessions that help find pathways that could address the digital divide issues. However, to date, there is little progress as we see the struggles that have led to the continuous mismanagement of the pandemic. The blame cannot be simply on people or governments but on lack of support and institutions that are meant to provide an inclusive society for all.

BOX 2.8 COVID-19 and Education

Educating the community has been a struggle since the start of the pandemic. Unfortunately, this has also been experienced in previous pandemic or epidemic events, showing a lack of framework or support in making people aware, enforcing compliance regulations, and implementing policies that could save people, businesses, and organisations.

The below examples are some selected cases of education from various viewpoints or perspectives.

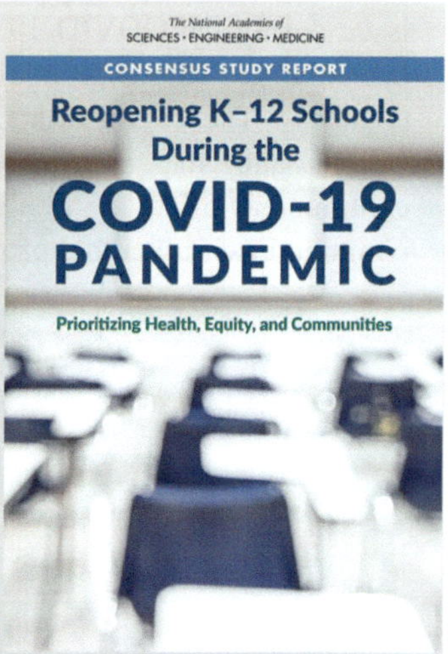

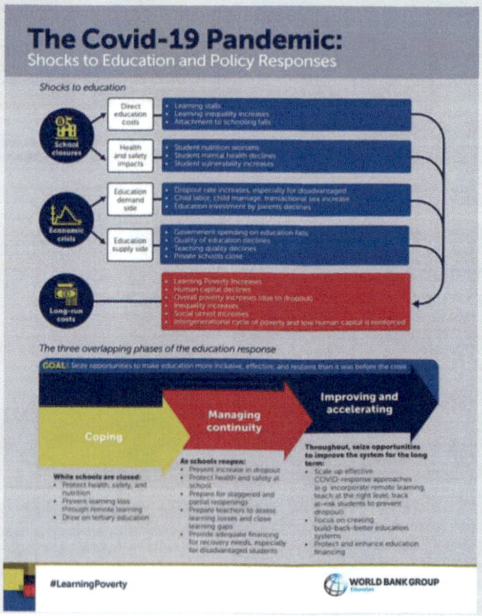

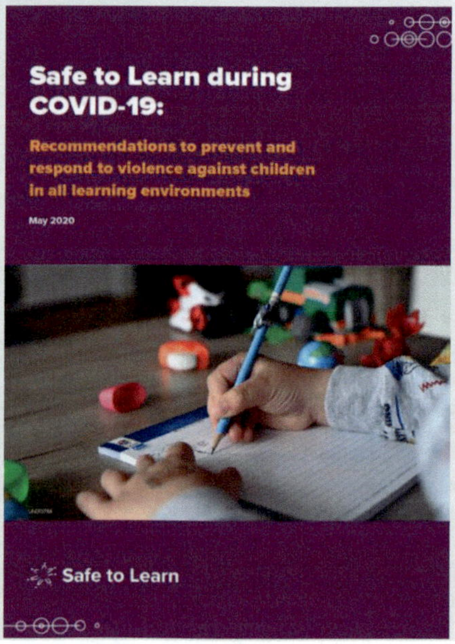

BOX 2.9 Transportation, Mobility, and People's Flow

To date, we still argue the most significant issue of pandemic continuity is the lack of control on transportation, mobility, and people's flow. Months after the start of the COVID-19 pandemic, the same issue persists. We see new hotspots are often places of high-level mobility and people's flow, where common practices of social distancing, careful monitory, and safety checks become more of the formality than applied routine.

Hongqiao Railway Station in Shanghai, as shown below (photograph taken in March 2021), is one of the main transportation hubs. During a busy weekday, amid the ongoing COVID-19 pandemic, we see a large influx of people going in and out of a shared indoor environment. Despite high-level control and safety measures, it is almost impossible to keep track of every mobility while maintaining normal daily operations. Even though Shanghai has been protected well so far, we see such places could simply turn into immediate hotspots. Therefore, we ought to reflect on this situation more cautiously to see what can be done to maintain a better balance for people's flow, intercity and intra-city mobility, and transportation between places.

BOX 2.10 Density, Urban Environments, and People

To date, the correlation between density and the spread of disease is mostly wrongly studied. In most cases, density is combined or interweaved with other negativities, which brings forward the bigger image of urban issues, such as poverty, hygiene, lack of frameworks, and lack of management. The examples below represent high-density (in top image) versus mixed-use compact development (in below image). The two are very different, and studies that explore density–COVID-19 nexus do not explore location-based data vs. real data for the disease spread. Unfortunately, such thinking may still suggest high density is acceptable while neglecting the many issues that it could bring if compactness is not understood correctly.

BOX 2.11 Urban Patterns and Living Environments

Urban patterns and physical forms of cities and communities play a significant part in making resilient environments. Yet, we often neglect urban design values, thinking that we can get smarter by just implementing smart everything in smart cities, communities, and places.

Later in the book, the topic of urban form is explored in detail. This is an important factor as part of the physical aspect of cities and communities. The way cities are planned and designed helps us achieve better resolutions for healthier communities, urban layouts, and inclusive societies.

The image here is from the central part of Chongqing in China, where the city is recognised well for its COVID-19 management plan from the early days of the pandemic.

An ideal urban form is what promotes health and sustainability. But, this cannot be a standard model for all.

Unfortunately, not all people have access to such public inner green environments (as shown in this image). Also, not all can consider high-rise development to be a suitable living environment. Indeed, we may just need to compromise.

BOX 2.12: Intercity and Intra-city Public Transportation

A neglected area that needs further study is the impact of the COVID-19 on intercity and intra-city public transportation and vice versa. In larger cities, we see growing issues related to intercity public transportation demand. As shown below, in cities like Beijing, there is a significant decline in the use of buses in cities. The situation has led to a growing demand for private car use, which itself could reverse many efforts of investing in public transportation projects/programmes.

The intra-city public transportation has a different story, and we see different issues arising. Here, we have two sides of the coin, one that entails issues related to lack of control and management of public transportation services. The other includes issues related to privacy and mobility control.

Many regions have limited strategies in defining boundaries between cities and their neighbouring cities/towns/villages. With this standard model in many places, intra-city mobility, through both public and private transportation means, could not be easily monitored or controlled. Yet, we see a growing demand in integrated models of intra-city mobility checks, such as digital tickets and integrated ID/passport and ticket purchases.

BOX 2.13: The Accelerating Digitisation during and after the COVID-19

The fast emergence of digitalisation and digitisation is surrounding us before we know it. In the name of smartness, we will see a growing demand for implementing and integrating smart technologies, digital apps, etc., in our everyday environments. We have to consider these with both positive and negative views.

Below are two examples of digital use for daily operations. On the left, we see an example of patient registration in a local hospital, and on the right, we see an example of a free facial mask collection point at a metro station. The COVID-19 pandemic has partly enabled this shift.

Here is what has been covered in one of our earlier works (see Cheshmehzangi 2021a):

> In a short time, during the early phases of the COVID-19 pandemic outbreak, we managed to shift rapidly to use digital technologies and replace some of our daily operations with virtual modes. This shift happened so instantly and widely that it enables us to argue that the COVID-19 became a valid reason to boost some of the gradual and ongoing transitions towards faster transformations.

BOX 2.14: The Changes that are "No Transitions", but "Transformations"

"*In all contexts, regardless of the technological advancement levels, the use of technologies in managing COVID-19 and daily operations has been a tangible part of these recent months. There is no doubt that COVID-19 has become a drive to boost the use of technologies in all sectors. From the increasing popularity in e-commerce to the use of virtual and digital platforms for various purposes, new apps, and new communication modes, we see a growing demand for what could be a new chapter in the development of our systems. The built environment sector, being part of this development, would possibly utilize the available toolkits even further and push for new policy formation based on the use of digital technologies*" (see Cheshmehzangi 2021a).

In below, we have two examples of AI facial recognition integration into a shopping mall entrance (on the left). The device is integrated into a temperature detection gate and works with a camera as well. On the right, we see the same applies to added steps for airport arrivals. In this example, a new channel is introduced, which is equipped with a facial recognition device, temperature control, and QR health check system. At the end of the route, an official person sits to take records and checks the documents.

"*The speeding digitalization…is expected to boost some of the earlier initiatives such as digital agenda, digital innovation in design, cyber city and cybersecurity, smart built environment, virtual education, and so on. After the current adversities of the pandemic are over, the chance for collaborative digital platforms will boost to a new level that includes more of detection and tracking tools, facial recognition technology, AI applications, future IoT, and digital security systems. In name of safety and security, some of these platforms will be pushed forwarded faster and are expected to be accepted by the general public sooner than anticipated. The role of the built environment will not be minimal as new paradigms are yet to emerge in the near future*" (see Cheshmehzangi 2021a).

BOX 2.15: A Growing Digital World: Issues of Digital Divide and Digital Everything

As shown in below images, the issues of the digital divide are recognised by various stakeholders, too. The help is needed for senior members, people with special needs, people without smart mobile phones, people without specific apps/applications, people who cannot register on the system, and even non-local passengers. The two examples are from two different airports in China during the months of April and May 2021.

The health QR code is widely used in different countries, like China. However, each city or region has adopted the model and creates its own platform. This causes inconsistency, but also some divide for users. The data is still stored centrally, which could cause other safety and security issues.

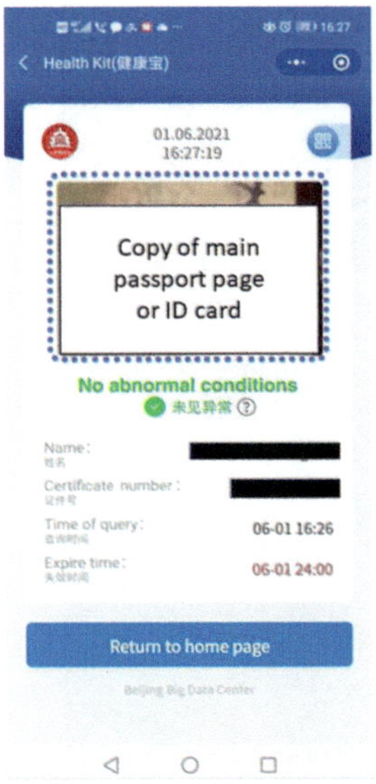

BOX 2.16 Emerging Digital and Track and Record

The digital use appearance is everywhere around us, from the food ordering in an airplane, to food delivery system and routine checks, and daily vegetable and fruit purchases.

Digital recording has become a common norm, from digital payment platforms to digital registration, recording, and inserting detailed information. Such

information is usually recorded and could be tracked when and if needed. This is the reason behind some of the recent democratic movements against digitalisation, privacy issues, and track and record platforms.

Below is an example of the track and record process at an airport arrival hall in April 2021.

BOX 2.17: Digital and "Track and Record"

Privacy issues are highlighted in track and record processes, and some countries (like Japan) have adopted a system that aims to protect personal information.

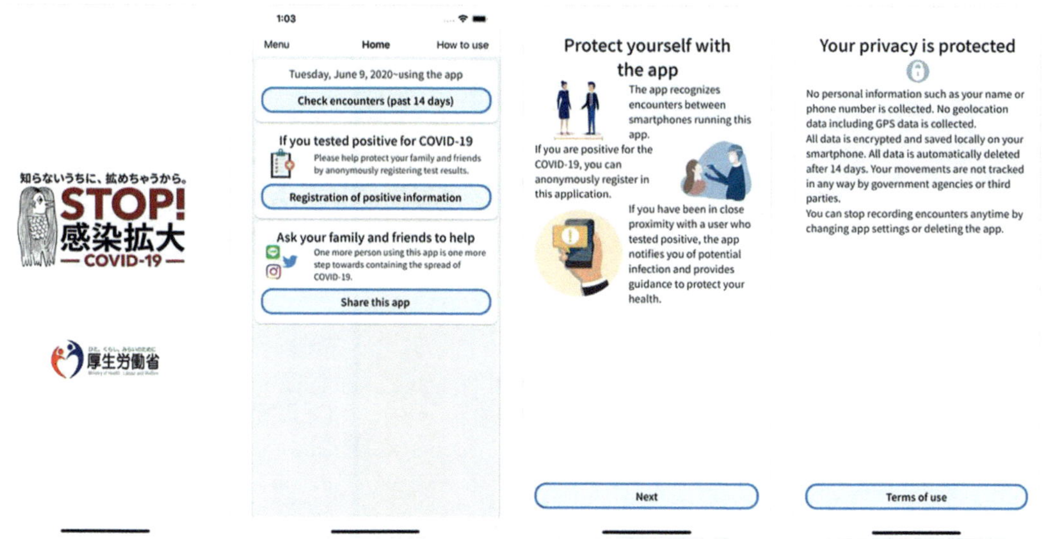

The UK's NHS system has also developed a comprehensive digital platform for all users. It is aimed to help stop the spread of the disease across the country.

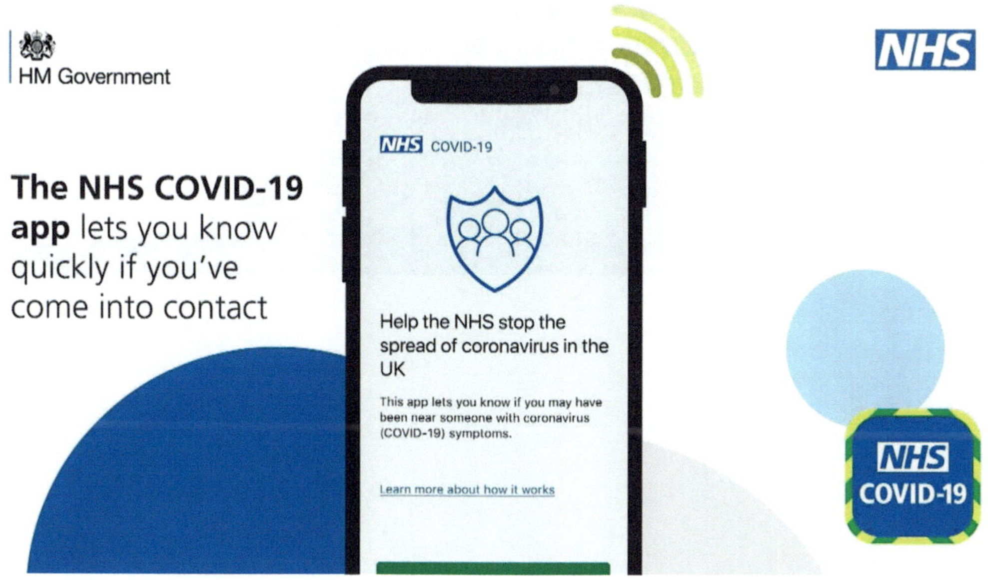

BOX 2.18 Digital and "Track and Record"

The digital move to include the COVID-19 vaccinated groups and distinguish them digitally has become a common norm in many countries. In China, the golden frame around the green code means the person is vaccinated. In other countries, like in the European Union (EU), the approach is also to create a digital version of the vaccination certificate. There is a push against the idea of vaccination passport in many countries, but will that be a common norm? Where are the limits in the name of security?

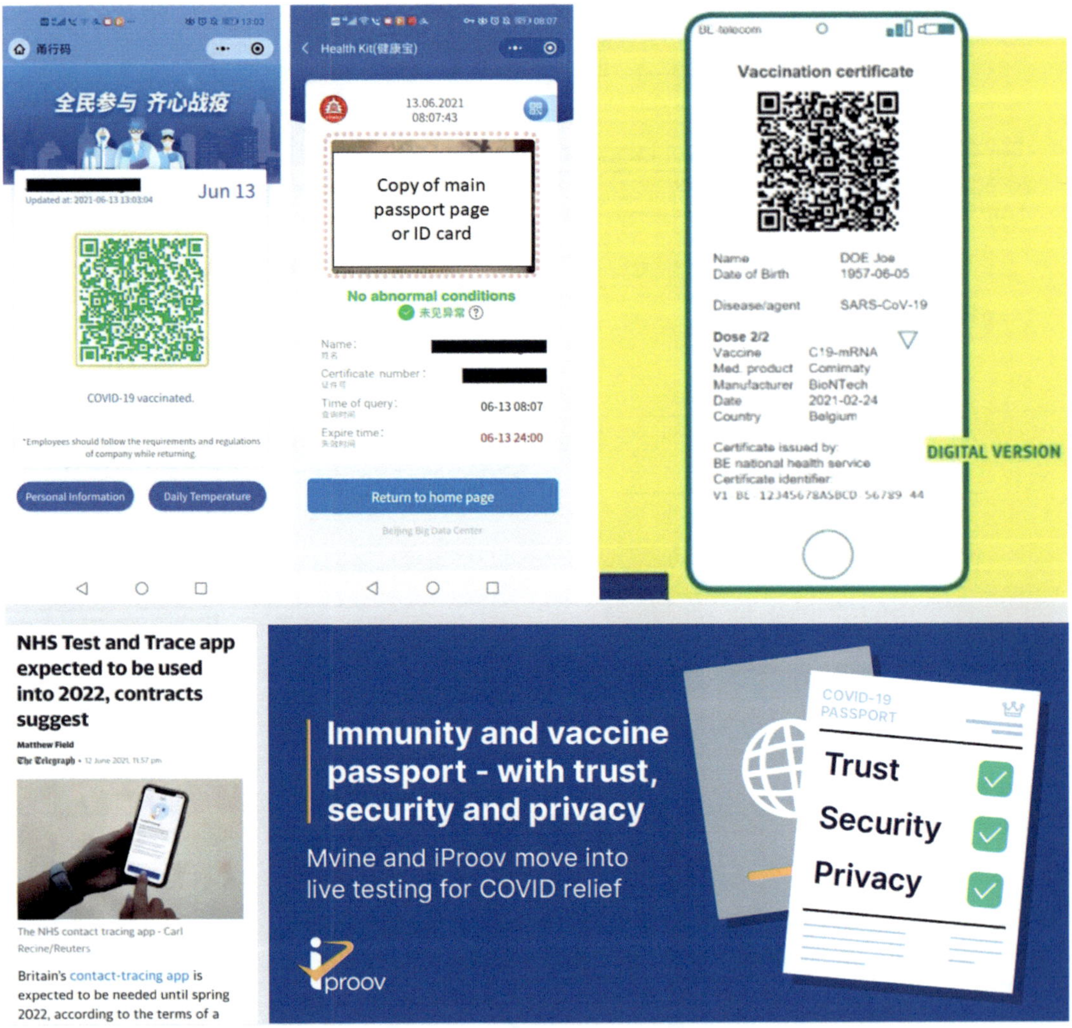

BOX 2.19: Digital Apps for Tracing and Health Checks

The move towards using tracing and health checking has become a common practice in many countries. The many localised digital applications, such as those in Singapore and the USA (as shown below), are good examples of such platforms that are used by the public. Some of these applications are owned by the governments or governmental/state-owned agencies, and some are owned by private enterprises or associated organisations.

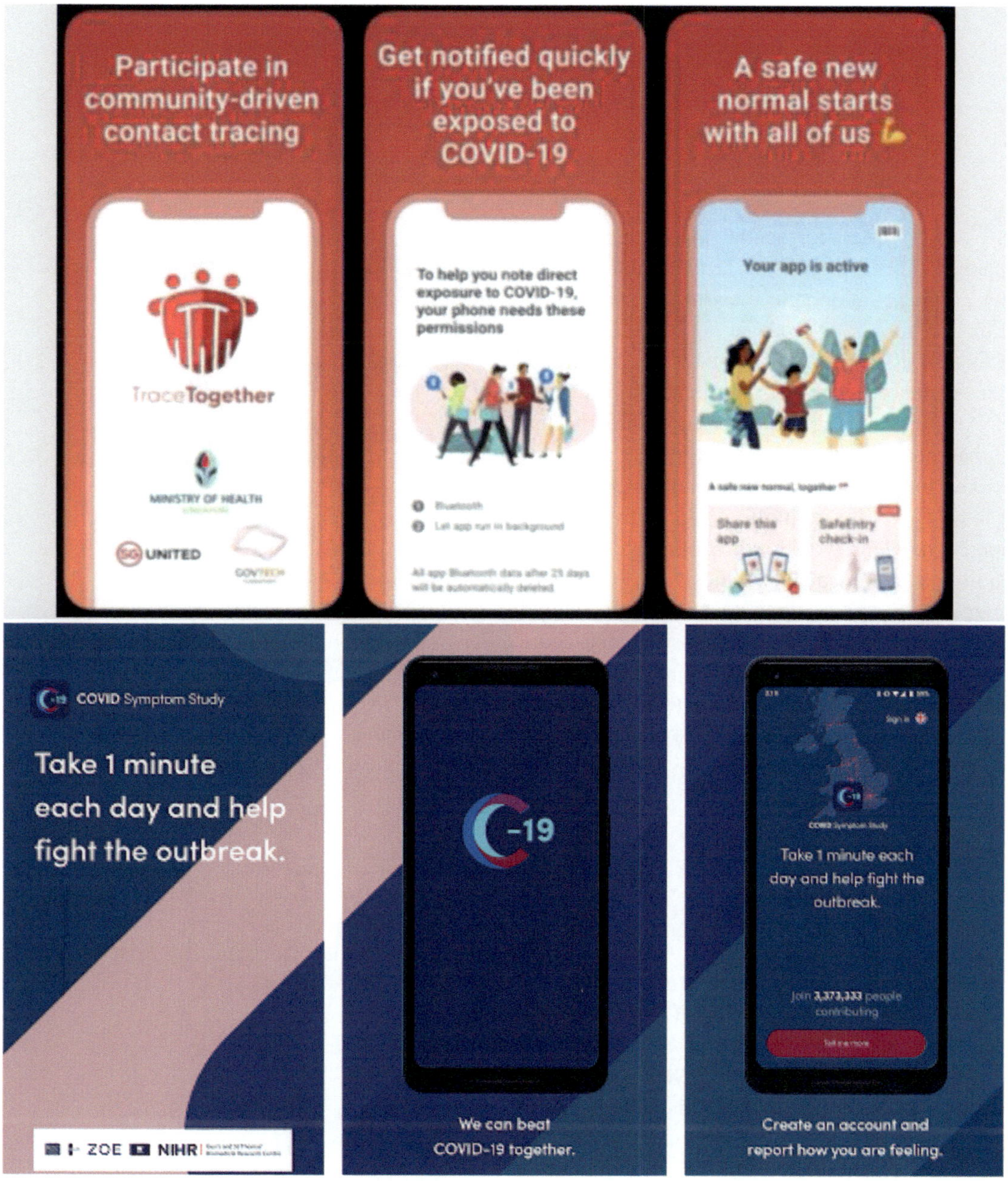

BOX 2.20: Public Realms: Reflecting on Equality, Accessibility, and Formality Issues

Here are some of our last reflections on equality, accessibility, and formality issues. These are partly covered in the following chapters (particularly related to physical and social dimensions), while we aim to stress more on inequalities, injustice, and poverty matters. Under the public realm discussions, we see examples of access to greenery and green spaces (top image), and the formality of processes such as temperature checks, space closures, control, and management (below image).

References

Abedi V, Olulana O, Avula V, Chaudhary D, Khan A, Shahjouei S, Zand R (2020) Racial, economic, and health inequality and COVID-19 infection in the United States. J Racial Ethnic Health Disparities 1–11

Ahmed F, Ahmed NE, Pissarides C, Stiglitz J (2020) Why inequality could spread COVID-19. Lancet Publ Health 5(5):e240

Amadei B (2020) Revisiting positive peace using systems tools. Technol Forecast Social Change 158:120149. https://doi.org/10.1016/j.techfore.2020.120149

Amadei B (2021). A systems approach to the sustainability-peace nexus. Sustain Sci 1–14. https://doi.org/10.1007/s11625-020-00902-x

Arain S, Thalapparambath R, Al Ghamdi FH (2021) COVID-19 pandemic: response plan by the Johns Hopkins Aramco Healthcare inpatient pharmacy department. Res Social Admin Pharm 17(1):2009–2011

Bambra C, Riordan R, Ford J, Matthews F (2020) The COVID-19 pandemic and health inequalities. J Epidemiol Community Health 74(11):964–968

Barash DP (2017) Approaches to peace, vol 199. Oxford University Press, Oxford

Beaunoyer E, Dupéré S, Guitton MJ (2020) COVID-19 and digital inequalities: reciprocal impacts and mitigation strategies. Comput Human Beh 111:106424

Betsch C, Korn L, Sprengholz P, Felgendreff L, Eitze S, Schmid P, Böhm R (2020) Social and behavioral consequences of mask policies during the COVID-19 pandemic. Proc Natl Acad Sci 117(36):21851–21853

Biswas RK, Huq S, Afiaz A, Khan HT (2020) A systematic assessment on COVID-19 preparedness and transition strategy in Bangladesh. J Eval Clin Pract 26(6):1599–1611

Blundell R, Costa Dias M, Joyce R, Xu X (2020) COVID-19 and Inequalities. Fisc Stud 41(2):291–319

Bond CJ (2014) Positive peace and sustainability in the mining context: beyond the triple bottom line. J Clean Prod 84:164–173

Bowleg L (2020) We're not all in this together: on COVID-19, intersectionality, and structural inequality. Am J Publ Health (AJPH). Available from: https://ajph.aphapublications.org/doi/full/https://doi.org/10.2105/AJPH.2020.305766

Brauch HGN, Oswald U, Grin J, Scheffran JR (2016) Handbook on sustainability transition and sustainable peace. Springer, Cham

Cahill-Ripley A (2016) Reclaiming the peacebuilding agenda: economic and social rights as a legal framework for building positive peace-a human security plus approach to peacebuilding. Hum Rights Law Rev 16(2):223–246

Cheshmehzangi A (2020b) The city in need: urban resilience and city management in disruptive disease outbreak events. Springer Singapore

Cheshmehzangi A (2020a) Housing and health evaluation related to general comfort and indoor thermal comfort satisfaction during the COVID-19 lockdown. J Human Behav Social Environ 1–26. https://doi.org/10.1080/10911359.2020.1817225

Cheshmehzangi A (2021b) Revisiting the built environment: 10 potential development changes and paradigm shifts due to COVID-19. J Urban Manage (In Press). https://doi.org/10.1016/j.jum.2021.01.002

Cheshmehzangi A (2021) From transitions to transformation: a brief review of the potential impacts of COVID-19 on boosting digitization, digitalization, and systems thinking in the built environment. J Build Constr Plan Res 9(01):26

Clark JN (2009) From negative to positive peace: the case of Bosnia and Hercegovina. J Human Rights 8 (4):360–384

Clune WH, Zehnder AJ (2020) The evolution of sustainability models, from descriptive, to strategic, to the three pillars framework for applied solutions. Sustain Sci 1–6

Collaborative C (2020) Global guidance for surgical care during the COVID-19 pandemic. Brit J Surg 107(9), 1097–1103. Available from: https://www.ncbi.nlm.nih.gov/pmc/articles/PMC7262310/

Dang HAH, Nguyen CV (2021) Gender inequality during the COVID-19 pandemic: income, expenditure, savings, and job loss. World Develop 140:105296

de Leeuw E, Green G (2017) The logic of method for evaluating healthy cities. In: de Leeuw E, Simos J (eds) Healthy cities: the theory, policy, and practice of value-based urban planning. Springer, New York, pp 463–477

Dell'Ariccia G, Mauro P, Spilimbergo A, Zettelmeyer J (2020) Economic policies for the COVID-19 war. IMF Blog 1

Desai AN, Aronoff DM (2020) Food safety and COVID-19. JAMA 323(19):1982–1982

Elgar FJ, Stefaniak A, Wohl MJ (2020) The trouble with trust: time-series analysis of social capital, income inequality, and COVID-19 deaths in 84 countries. Social Sci Med 263:113365

Engzell P, Frey A, Verhagen MD (2020) Learning inequality during the COVID-19 pandemic. Available from: https://osf.io/download/5f995b4687b7df03233b06fe/

Fischer D (2007) Peace as a self-regulating process. In Webel C, Galtung J (eds) Handbook of peace and conflict studies, Chap. 13. Routledge, New York

Fisher AN, Ryan MK (2021) Gender inequalities during COVID-19. Group Process Intergroup Relat 24 (2):237–245

Fisher J, Arora P, Chen S, Rhee S, Blaine T, Simangan D (2021) Four propositions on integrated sustainability: toward a theoretical framework to understand the environment, peace, and sustainability nexus. Sustain Sci 1–21. https://doi.org/10.1007/s11625-021-00925-y

Galtung J (1964) An editorial. J Peace Res 1(1):4

Galtung J (1969) Violence, peace, and peace research. J Peace Res 6(3):167–191

Galtung J (1990) Cultural violence. J Peace Res 27(3): 291–305

Galtung J (2011) Peace, positive and negative. In: The Encyclopedia of peace psychology. Available from: https://doi.org/10.1002/9780470672532.wbepp189

Galtung J, Fischer D (2013) Positive and negative peace. In Johan Galtung. Springer, Berlin, Heidelberg, pp 173–178

Grewal BS (2003) Johan Galtung: positive and negative peace. School Social Sci Auckland Univ Technol 30:23–26

Ho KFW, Ho KF, Wong SY, Cheung AW, Yeoh E (2020) Workplace safety and coronavirus disease (COVID-19) pandemic: survey of employees. Bulletin World Health Org. Available from: https://www.who.int/bulletin/online_first/20-255893.pdf

Hsiang S, Allen D, Annan-Phan S, Bell K, Bolliger I, Chong T, Druckenmiller H, Huang LY, Hultgren A, Krasovich E, Lau P (2020) The effect of large-scale anti-contagion policies on the COVID-19 pandemic. Nature 584(7820):262–267

IEP (Institute for Economics & Peace) (2018) Global peace index 2018: measuring peace in a complex world. Available from: http://visionofhumanity.org/reports

IEP (Institute for Economics & Peace) (2019a) Positive peace report 2019: analysing the factors that sustain peace. Institute for Economics & Peace, Sydney

IEP (Institute for Economics & Peace) (2019b) Global peace index 2019: measuring peace in a complex world. Institute for Economics & Peace, Sydney

IEP (Institute for Economics & Peace) (2017) Positive peace report 2017: tracking peace transitions through a systems thinking approach. Report number 54, IEP, Sydney, Australia

Kidd CV (1992) The evolution of sustainability. J Agric Environ Ethics 5(1):1–26

Kim SJ, Bostwick W (2020) <covid19> Social vulnerability and racial inequality in COVID-19 deaths in Chicago. Health Educ Behav 47(4):509–513

Milesi C, Franco EL (2020) Volunteer contributions to building more peaceful, inclusive, just and accountable societies. Available from: http://ceciliamilesi.com/wp-content/uploads/2020/09/Paper_Volunteer-Contributions_SDG16Plus-1.pdf

Mossa-Basha M, Medverd J, Linnau KF, Lynch JB, Wener MH, Kicska G, Sahani DV (2020) Policies and guidelines for COVID-19 preparedness: experiences from the University of Washington. Radiology 296(2): E26–E31

Okoi O, Bwawa T (2020) How health inequality affect responses to the COVID-19 pandemic in Sub-Saharan Africa. World Devel 135:105067

Patel JA, Nielsen FBH, Badiani AA, Assi S, Unadkat VA, Patel B, Ravindrane R, Wardle H (2020) Poverty, inequality and COVID-19: the forgotten vulnerable. Publ Health 183:110

Piguillem F, Shi L (2020) Optimal COVID-19 quarantine and testing policies. CEPR discussion paper, No. DP14613, 42 pages. Available from: https://papers.ssrn.com/sol3/papers.cfm?abstract_id=3594243

Planning Tank (2017) Healthy city: implication of public health and environment protection. Available from: https://planningtank.com/urbanisation/healthy-city-public-health-environment

Positive Peace (n.d.) What is positive peace? Our understanding of the term 'peace' has evolved significantly over the last 2500 years, Available from: https://positivepeace.org/what-is-positive-peace

Reardon BA (1988) Comprehensive peace education: ducating for global responsibility. Teachers College Press, 1234 Amsterdam Avenue, New York, NY 10027

Roberts D (2008) Post-conflict statebuilding and state legitimacy: from negative to positive peace? Dev Chang 39(4):537–555

Rodriguez-Morales AJ, Gallego V, Escalera-Antezana JP, Méndez CA, Zambrano LI, Franco-Paredes C, Suárez JA, Rodriguez-Enciso HD, Balbin-Ramon GJ, Savio-Larriera E, Risquez A (2020) COVID-19 in Latin America: the implications of the first confirmed case in Brazil. Travel Med Infect Dis 35:101613. Available from: https://www.ncbi.nlm.nih.gov/pmc/articles/PMC7129040/

Sharifi A, Simangan D, Kaneko S (2021) The literature landscape on peace–sustainability nexus: a scientometric analysis. Ambio 50:661–678. https://doi.org/10.1007/s13280-020-01388-8

Sharp DN (2011) Addressing economic violence in times of transition: towards a positive-peace paradigm for transitional justice. Fordham Int LJ 35:780

Shields P (2017) Special commentary: limits of negative peace, faces of positive peace. Parameters 47(3):5–14. Available from: https://digital.library.txstate.edu/handle/10877/6980

Su Z, McDonnell D, Wen J, Kozak M, Abbas J, Šegalo S, Li X, Ahmad J, Cheshmehzangi A, Cai Y, Yang L (2021) Mental health consequences of COVID-19 media coverage: the need for effective crisis communication practices. Glob Health 17(1):1–8

United Nations (UN) (2019) Global Sustainable development report 2019: the future is now—science for achieving sustainable development, United Nations, New York. Available from: https://sustainabledevelopment.un.org/globalsdreport/2019

Van Dorn A, Cooney RE, Sabin ML (2020) COVID-19 exacerbating inequalities in the US. Lancet (london, England) 395(10232):1243

Virji H, Sharifi A, Kaneko S, Simangan D (2019) The sustainability–peace nexus in the context of global change. Sustain Sci 14:1467–1468. https://doi.org/10.1007/s11625-019-00737-1

World Health Organization (2020) Critical preparedness, readiness and response actions for COVID-19: interim guidance, 22 March 2020 (No. WHO/2019-nCoV/Community_Actions/2020.3). World Health Organization

Part I
Sustainability and Peace: "Urban Health from the Physical Dimension"

In the first part of studying sustainability and peace, we present three primary aspects representing urban health from the physical dimension.

The first aspect is urban density and land use, exploring the health-related matters that should be considered from the planning and design perspectives. These are discussed in Chap. 3, focusing on keeping the peace and security of the built environment through better planning and advocating compact design development but with some key considerations. The second aspect is urban morphology and form, highlighting an overarching area of resilient urban form, specifically discussing the spatio-morphological aspect of cities and communities. These are discussed in Chap. 4, explicitly exploring the health and safety of communities from the urban form perspective. The third aspect is critical infrastructures, including a range of critical physical environments and uses in cities, suggesting their role in sustaining the peace in our communities. This important aspect is presented in Chap. 5, identifying the importance of critical infrastructures in achieving urban health, sustainability, and peace.

The Role of Urban Density and Land Use Planning in Keeping the Cities Healthy

3

The self is always subject to its context.
—Ali Cheshmehzangi, 2012

3.1 A General Overview of Urban Density and Land Use Planning

A major driver in achieving healthy cities would be to consider a balanced urban density in cities, which could then be supported with adequate land use planning measures. In earlier studies, we looked into the relationship between sustainable living and urban density (Cheshmehzangi and Butters 2015, 2016), understanding living qualities in urban environments of different densities (Cheshmehzangi and Butter 2017a) and opposing the approaches that lead towards dense high-rise cities (Butters et al. 2020). In the earlier studies, we aimed to relate a variety of factors that are embedded in urban density functionalities (Mills 1970), implications on sustainable development (Roberts 2007), and other aspects more involved in population distribution and imbalance in city environments (Griffith 1981). Perhaps higher densities are economically viable (McDoland 1989), but there are issues that suggest higher poverty in denser living environments (Dodman 2009) and other socio-economic problems that may become issues of crowding or overcrowding (Baldassare 1979) rather than healthy urban densities. In most cases, we note that higher density and land use are interlinked, leading to difficulties in cities and towards unhealthy urban conditions, such as congestion (Wheaton 1998), land value increase (Bocarejo et al. 2013), imbalance in urban living areas (Shoshany and Goldshleger 2002), higher level of crime and violence in neighbourhood contexts (Browning et al. 2010), as well as health issues. The examples of health impacts from higher density include cases of higher social disorganisation in city environments (Schmitt 1966), psychological health (Evans et al. 1989, 2001), sanitation (Hathi et al. 2017; Cheshmehzangi 2020), etc. The relationship of urban design to human health and condition (Jackson 2003) has already proven to be a major part of how cities should be developed, rather than densification matters that lead to unhealthy urban conditions and poorer communities. In this regard, we can argue that peaceful living environments could be compact environments, but not necessarily of high-density nature.

In earlier work, we highlighted the difference between high-density living environments and compact urban design solutions (Butters et al. 2020). The latter involves healthier community setting allowing for better interaction opportunities, healthier urban greenspace environments (Kardan et al. 2015), and better resilience against issues of mental health (Wandersman and Nation 1998), which should minimise conflicts and

A. Cheshmehzangi, *Urban Health, Sustainability, and Peace in the Day the World Stopped*,
Sustainable Development Goals Series, https://doi.org/10.1007/978-981-16-4888-5_3

achieve healthier urban structures (Melis et al. 2015). In this chapter, we explore the first topic of the physical dimension, addressing what are the health implications (Lo and Quattrochi 2003) from unhealthy densities and land use planning. Some aspects are common sense, such as higher noise levels associated with urban densities and specific land use patterns (King et al. 2012), and some could be more relevant to the book's aim, such as matters of mobility and transportation (Cervero 2013; Cheshmehzangi 2020), public health and quality of life (Frank 2000), diversity matters (Lu et al. 2017; Cheshmehzangi and Li 2020), and environmental qualities (Shi et al. 2017) associated with the urban environments. For us, it is important to highlight the role of balanced urban density and land use planning in keeping cities healthy and viable.

3.2 Reflecting on How Cities Reacted During the Pandemic

The correlation between urban density and the COVID-19 spread cannot be neglected. Earlier studies indicate the link exists where higher population density leads to the spread and severity of the disease (Carozzi 2020), which again is statistically proven to be the case in other contexts (Kadi and Khelfaoui 2020; Bhadra et al. 2021). While density could be regarded as a positive matter in regards to having more health centres and better public services (but not in all cases), it could also be seen as "*the intrinsic capacity of a city to implement preventive physical segregation policies in public transport, public spaces, as well as shared services and facilities*" (Lai et al. 2020a, b). The ongoing pandemic points out a higher rate of infections and death rates in contexts where higher density is mixed with more deficient land use planning and poorer communities. We see a much higher rate in populated and higher density communities in the USA, Europe, Latin American countries, India, China, Japan, etc. In China, for instance, more severe conditions were identified in urban settings of higher density populations (Ni et al. 2021). In low-income countries (LICs), this also became a major problem resulting in critical epidemic situations such as intense social mixing (Favas et al. 2020), uncontrollable situations, inadequate access to basic infrastructure services (Sharifi and Khavarian-Garmsir, 2020), and other aspects that led to unhealthy living environments (Cheshmehzangi 2021). In other words, the healthiness of city became more fragile in the context where urbanisation and urban development are not controlled, land use mixes are not planned adequately, and densities are very high and combined with other issues of poorer living qualities and sanitary conditions (Sharifi and Khavarian-Garmsir 2020). The situation has also enabled us to question more carefully the sustainability of larger cities, such as megacities (Desai 2020), and consider the role of the urban settlement system in the spread and control of the pandemic (Gargiulo et al. 2020). Undoubtedly, these factors become ever important for us to revisit the built environments, particularly from the consideration of density and land use planning.

Here, we have two sides to the coin. On one side, we see positive aspects such as those related to better facilities and healthcare systems, both quantitatively and qualitatively. On the other side, we see negativities relating to lower quality living environments, often defined as dense slums or poorer communities that are unhealthy, lack sanitation, and face difficulties in maintaining healthy living environments (Cheshmehzangi and Butters 2017b). More importantly, we see common issues in both developed and developing contexts but in different forms. In both contexts, higher density and larger-scale cities are known as the major economic hubs of the respective countries or regions. Hence, we see higher population mobility and higher risk for the spread of disease outbreaks like the COVID-19. In both contexts, cities and communities of such nature suffered equally and differently at the same time. Equally, they both experienced higher rates of infection and death rates. Differently, one faced poorer conditions of healthcare systems, control, and monitory, while the other could cope with the situation with better services and support. However, our argument remains that higher density without the consideration of compact community layouts, adequate land use planning, and density balance could lead to major disruptive issues, creating unhealthy living environments beyond the control.

In the case of the COVID-19 original epicentre, as per records, we see a higher distribution of COVID-19 morbidity rate in correlation with key social and economic factors (You et al. 2020), showing that findings of earlier studies on density are not entirely accurate (Hamidi et al. 2020). While Hamidi et al. (2020) suggest connectivity is more important than density, we note the two are strongly interrelated. Contexts, where urban density is higher, would by default have other implications of higher connectivity, higher mobility, and a higher level of social interactions. For instance, some studies highlighted the issues of density-based clustering in cities (Li et al. 2021), indicating that urban mobility and density are indeed interrelated and could cause higher risks in higher populated and denser communities. The issues have been persistent in contexts that have faced several waves of the COVID-19 pandemic, such as in Japan (Arimura et al. 2020). The complexity of socio-spatial implications of COVID-19 measures, as Salama (2020) highlights, is an important determinant in cities. Nonetheless, it is very important to differentiate between compactness and high density. Mixing the two could lead to dangerous decision-making procedures in planning guidelines, leading towards even unhealthier living environments such as sprawled and dispersed cities.

When we first started studying density-COVID-19 correlations in summer 2020, we looked into the situation from the city-level perspective. The comparison study then showed us deficiency in findings, proving that city-level density comparisons could conflict with other factors that could simply be neglected in the correlations. Therefore, we noticed it is more applicable to study district level or neighbourhood levels to clarify better how density and disease spread and control are correlated. As highlighted by Acuto (2020), upward high-rise density has become a common characteristic of urban centres globally, showing that this trend could also lead to issues that may be more complicated by the traits of contemporary urbanism. We also recognise the role of movement and mobility in the way cities reacted or acted during the pandemic (Mitra et al. 2020), suggesting that specific (urban) spatial features could lead to contextual issues that may be unique to particular regions and cities. The current situation inspires us to move towards better improvements (Megahed and Ghoneim 2020), considering indeed the almost never-ending issues of urban health in an informal settlement (Wilkinson 2020), control factors in cities and communities (de Oliveira et al. 2020), and understanding adaptive approaches and measures (Teller 2021) to find positive and negative associations between density and other urban factors. Living in crowded environments could be problematic for any city of any status but should also be studied in specific contexts and how crowding could have consequences (von Seidlein et al. 2020), which are impactful to urban environments (Lee et al. 2021). By considering the urban nature (Venter et al. 2020) in the event of similar crises, we note the importance of reflecting on urban density and land use planning to ensure health and peace are sustained.

3.3 Addressing the Flaws

The answer to what has been identified as the primary flaw is the term "balance". We often see the complexity of urban parameters (Pirouz et al. 2020) as an overarching situation to consider and measure the resiliency of cities and communities. For us, it is important to identify vulnerabilities (Mishra et al. 2020) and inequalities that exist between different communities, perhaps in the same city or in different contexts. We cannot neglect the importance of changing dynamics during the pandemic (Iranmanesh ad Alpar Atun 2021), which led to a considerable understanding of city operations and management (Cheshmehzangi 2020). By addressing the flaws of high urban density and improper land use planning, we would be able to tackle the significant issues that impact the healthiness and well-being of communities. As suggested in the findings of Sahasranaman and Jensen (2021), "*neighbourhood population density is a critical mediator of the dynamics of infectious spread in a city*", particularly when combined with other factors of poverty and socio-economic inequalities. For us to reflect well on the current situation, we see an opportunity to revise social and environmental factors that have led to the faster spread of diseases in communities that face inequality and have a high population of poverty (Nakada and Urba 2020).

By questioning whether the city is the core of contagion, Biglieri et al. (2020) evaluate the social and spatial aspects of urban society. In such cases, we see a great opportunity to evaluate how public health is maintained in unhealthy communities from socio-environmental factors and towards key factors of urban poverty, slum development, poorer social and health determinants, weaker infrastructures, etc. While many urban planning and design lessons are given in this prolonged pandemic (Pandit 2020), we see conflicts and unhealthy situations for more impoverished communities often dense in accommodating migrant workers and with higher risks. Therefore, it is only right to reconsider the built environment parameters that define the healthiness and well-being of the city from the perspectives of urban density and land use planning. In doing so, we are able to investigate and suggest better quality urban environments, including healthier community layouts, a more balance pattern for urban environments, and still consider the value of compact development. Once again, we oppose the high-rise high-density urban pattern (Butters et al, 2020), which has brought many negativities that are almost irreversible. These would indeed be relevant to other factors that would be studied in the following few chapters. The conditions of the high density and poor communities should be prioritised in developing peace and well-being that is essential for their healthy progression. If we fail to do so, not only that the ultimate Sustainable Development Goals (SDGs) are not achieved, but also we face widening inequalities that eventually divide our cities further. This division will be unhealthy for peace-making strategies in societies where the general public perceives and experiences more inequalities. If we fail to reverse such divisions, we may face further impacts on the overarching directions of inclusiveness and fairness (Gupte and Mitlin, 2020) that are the backbone of any healthy city and/or community.

3.4 A Summary

In most cases, high density alone is not a major issue as we have many cities of high density where they managed to contain and control the COVID-19 outbreak quite fast. A good example is Macao SAR, in south China, where it has one of the highest densities in the world. In this regard, we can argue that high density becomes an issue only in correlation with other factors—i.e. both positive and negative factors. The trendy example of high density and connectivity, mainly based on intra-city, regional, and international connectivity, is of major concern. Higher densities with a higher level of outward connectivity patterns are generally vulnerable in dealing with pandemics. Another example is high density with unstructured land use planning or low-quality urban design. In addition, the combination of high density and poverty could lead to significant urban challenges for residents and governments. This example is visible in cases of informal settlements and overcrowded slums, where many vulnerabilities exist and are generally overlooked.

To summarise briefly, we suggest revisiting densification strategies that have become the common global trends in urban planning and design. The careful consideration of land use planning with a balanced urban density could lead to better urban design solutions, advocating peace and urban health, particularly in communities that face longer-term adversities of poverty and decline. By considering the physicality of the urban environment, we could delve into prescriptive placemaking strategies to better define the spatial organisation of the city (Scott 2020) and inform policies and practices in urbanism (Lai et al. 2020a, b; Sallis et al. 2020). The improvements in land use planning and administration (Chigbu 2020) are essential to reflect on necessary urban transitions, focusing on promoting urban health and achieving societal well-being. One of the areas that would be beneficial for further exploration, specifically from the urban density and land use planning perspectives, would be to consider urban planning interventions (Bolleter et al. 2021; Chigbu and Onyebueke 2021), which would ultimately lead to the enhancement of the physical dimension of the built environment. To follow these further, the next chapter focuses on urban form and morphological aspects, which are relevant to the development of resilient and healthy communities from the physical dimension.

References

Acuto M (2020) COVID-19: Lessons for an urban (izing) world. One Earth 2(4):317–319

Arimura M, Ha TV, Okumura K, Asada T (2020) Changes in urban mobility in Sapporo city, Japan due to the Covid-19 emergency declarations. Transp Res Interdisc Perspect 7:100212

Baldassare M (1979) Residential crowding in urban America. University of California Press, Oakland

Bhadra A, Mukherjee A, Sarkar K (2021) Impact of population density on Covid-19 infected and mortality rate in India. Model Earth Syst Environ 7(1):623–629

Biglieri S, De Vidovich L, Keil R (2020) City as the core of contagion? Repositioning COVID-19 at the social and spatial periphery of urban society. Cities Health, pp 1–3

Bocarejo JP, Portilla I, Pérez MA (2013) Impact of Transmilenio on density, land use, and land value in Bogotá. Res Transp Econ 40(1):78–86

Bolleter J, Edwards N, Cameron R, Duckworth A, Freestone R, Foster S, Hooper P (2021) Implications of the Covid-19 Pandemic: Canvassing Opinion from Planning Professionals. Plan Pract Res, pp 1–22

Browning CR, Byron RA, Calder CA, Krivo LJ, Kwan MP, Lee JY, Peterson RD (2010) Commercial density, residential concentration, and crime: land use patterns and violence in neighborhood context. J Res Crime Delinq 47(3):329–357

Butters C, Cheshmehzangi A, Sassi P (2020) Cities, energy and climate: seven reasons to question the dense high-rise city. J Green Build 15(3):197–214

Carozzi F (2020) Urban density and COVID-19. IZA Discussion Paper No. 13440, 29 Pages, Available from: https://papers.ssrn.com/sol3/papers.cfm?abstract_id=3643204

Cervero R (2013) Linking urban transport and land use in developing countries. J Transp Land Use 6(1):7–24

Cheshmehzangi A, Butters C (eds) (2017) Designing cooler cities: energy, cooling and urban form—the Asian perspective. Palgrave Macmillan, Singapore

Cheshmehzangi A (2020) The city in need: urban resilience and city management in disruptive disease outbreak events. Springer, Singapore

Cheshmehzangi A, Butters C (2016) Sustainable living and urban density: the choices are wide open. Energy Procedia 88:63–70

Cheshmehzangi A, Butters C (2017) Chinese urban residential blocks: towards improved environmental and living qualities. Urban Des Int 22(3):219–235

Cheshmehzangi A, Li HMA (2020) Innovation through urban diversity and achieving comprehensive sustainable urbanism from a community-oriented approach. Curr Urban Stud 8(02):222

Cheshmehzangi A, Butters C (2015) Refining the complex urban: The study of urban residential typologies for reduced future energy and climate impacts. In: Proceedings of 8th conference of the international forum on urbanism. Incheon, South Korea

Cheshmehzangi A (2021) Revisiting the built environment: 10 potential development changes and paradigm shifts due to COVID-19. J Urban Manage. Available from: https://www.sciencedirect.com/science/article/pii/S2226585621000054

Chigbu UE (2020) e-Tracking COVID-19 disruptions to the global development agenda on land. Int J Environ Sci Nat Resour 26(1):1–9

Chigbu UE, Onyebueke VU (2021) The COVID-19 pandemic in informal settlements: (re) considering urban planning interventions. Town Plan Rev 92 (1):115–121

de Oliveira MM, Fuller TL, Brasil P, Gabaglia CR, Nielsen-Saines K (2020) Controlling the COVID-19 pandemic in Brazil: a challenge of continental proportions. Nat Med 26(10):1505–1506

Desai D (2020) Urban densities and the Covid-19 pandemic: Upending the sustainability myth of global megacities. ORF Occas Paper 244(4):1–4

Dodman D (2009) Urban density and climate change. Anal Rev Interact Urban Growth Trends Environ Changes (1). Available from: https://www.academia.edu/download/13774580/dodman_paper.pdf

Evans GW, Palsane MN, Lepore SJ, Martin J (1989) Residential density and psychological health: the mediating effects of social support. J Pers Soc Psychol 57(6):994

Evans GW, Saegert S, Harris R (2001) Residential density and psychological health among children in low-income families. Environ Behav 33(2):165–180

Favas C, Checchi F, Waldman RJ (2020) Guidance for the prevention of COVID-19 infections among high-risk individuals in urban settings. Available from: https://www.lshtm.ac.uk/sites/default/files/2020-05/Guidance_shielding_urban_FV_200501.pdf

Frank LD (2000) Land use and transportation interaction: implications on public health and quality of life. J Plan Educ Res 20(1):6–22

Gargiulo C, Gaglione F, Guida C, Papa R, Zucaro F, Carpentieri G (2020) The role of the urban settlement system in the spread of Covid-19 pandemic. The Italian case. TeMA-J Land Use Mobility Environ, pp 189–212
Griffith DA (1981) Modelling urban population density in a multi-centered city. J Urban Econ 9(3):298–310
Gupte J, Mitlin D (2020) COVID-19: what is not being addressed. Environ Urban
Hamidi S, Sabouri S, Ewing R (2020) Does density aggravate the COVID-19 pandemic? Early findings and lessons for planners. J Am Plann Assoc 86 (4):495–509
Hathi P, Haque S, Pant L, Coffey D, Spears D (2017) Place and child health: the interaction of population density and sanitation in developing countries. Demography 54(1):337–360
Iranmanesh A, Alpar Atun R (2021) Reading the changing dynamic of urban social distances during the COVID-19 pandemic via Twitter. Eur Soc 23 (sup1):S872–S886
Jackson LE (2003) The relationship of urban design to human health and condition. Landsc Urban Plan 64 (4):191–200
Kardan O, Gozdyra P, Misic B, Moola F, Palmer LJ, Paus T, Berman MG (2015) Neighborhood greenspace and health in a large urban center. Sci Rep 5(1):1–14
King G, Roland-Mieszkowski M, Jason T, Rainham DG (2012) Noise levels associated with urban land use. J Urban Health 89(6):1017–1030
Lai KY, Webster C, Kumari S, Sarkar C (2020) The nature of cities and the Covid-19 pandemic. Current Opinion in Environmental Sustainability 46:27–31
Lai S, Leone F, Zoppi C (2020) Covid-19 and spatial planning. TeMA-J Land Use Mobility Environ, pp 231–246
Lee W, Kim H, Choi HM, Heo S, Fong KC, Yang J, Park C, Kim H, Bell ML (2021).Urban environments and COVID-19 in three Eastern states of the United States. Sci Total Environ, p 146334
Li B, Peng Y, He H, Wang M, Feng T (2021) Built environment and early infection of COVID-19 in urban districts: a case study of Huangzhou. Sustain Cities Soc 66:102685
Lo CP, Quattrochi DA (2003) Land-use and land-cover change, urban heat island phenomenon, and health implications. Photogramm Eng Remote Sens 69 (9):1053–1063
Lu Y, Xiao Y, Ye Y (2017) Urban density, diversity and design: Is more always better for walking? A study from Hong Kong. Prev Med 103:S99–S103
McDonald JF (1989) Econometric studies of urban population density: a survey. J Urban Econ 26:361–385
Megahed NA, Ghoneim EM (2020) Antivirus-built environment: Lessons learned from Covid-19 pandemic. Sustain Cities Soc 61:102350
Melis G, Gelormino E, Marra G, Ferracin E, Costa G (2015) The effects of the urban built environment on mental health: a cohort study in a large northern Italian city. Int J Environ Res Public Health 12(11):14898–14915
Mills ES (1970) Urban density functions. Urban Stud 7 (1):5–20
Mishra SV, Gayen A, Haque SM (2020) COVID-19 and urban vulnerability in India. Habitat Int 103:102230
Mitra R, Moore SA, Gillespie M, Faulkner G, Vanderloo LM, Chulak-Bozzer T, Rhodes RE, Brussoni M, Tremblay MS (2020) Healthy movement behaviours in children and youth during the COVID-19 pandemic: exploring the role of the neighbourhood environment. Health Place 65:102418
Nakada LYK, Urban RC (2020) COVID-19 pandemic: environmental and social factors influencing the spread of SARS-CoV-2 in São Paulo, Brazil. Environ Sci Pollut Res, pp 1–7
Ni Z, Lebowitz ER, Zou Z, Wang H, Liu H, Shrestha R, Zhang Q, Hu J, Yang S, Xu L, Wu J (2021) Response to the COVID-19 outbreak in urban settings in China. J Urban Health 98(1):41–52
Pandit N (2020) COVID-19: Urban density's poetic justice. Ecol Econ Soc INSEE J 3(2):13–18
Pirouz B, Shaffiee Haghshenas S, Pirouz B, Shaffiee Haghshenas S, Piro P (2020) Development of an assessment method for investigating the impact of climate and urban parameters in confirmed cases of COVID-19: a new challenge in sustainable development. Int J Environ Res Public Health 17(8):2801
Sahasranaman A, Jensen HJ (2021) Spread of COVID-19 in urban neighbourhoods and slums of the developing world. J R Soc Interface 18(174):20200599
Salama AM (2020) Coronavirus questions that will not go away: interrogating urban and socio-spatial implications of COVID-19 measures. Emerald Open Res 2. Available from: https://www.ncbi.nlm.nih.gov/pmc/articles/PMC7219560/
Sallis JF, Adlakha D, Oyeyemi A, Salvo D (2020) An international physical activity and public health research agenda to inform COVID-19 policies and practices. J Sport Health Sci. Available from: https://www.ncbi.nlm.nih.gov/pmc/articles/PMC7243764/
Schmitt RC (1966) Density, health, and social disorganization. J Am Inst Plann 32(1):38–40
Scott M (2020) Covid-19, place-making and health. Available from: https://www.tandfonline.com/doi/full/https://doi.org/10.1080/14649357.2020.1781445
Sharifi A, Khavarian-Garmsir AR (2020) The COVID-19 pandemic: Impacts on cities and major lessons for urban planning, design, and management. Sci Total Environ, p 142391
Shi Y, Lau KKL, Ng E (2017) Incorporating wind availability into land use regression modelling of air quality in mountainous high-density urban environment. Environ Res 157:17–29
Shoshany M, Goldshleger N (2002) Land-use and population density changes in Israel—1950 to 1990: analysis of regional and local trends. Land Use Policy 19(2):123–133

Teller J (2021) Urban density and Covid-19: towards an adaptive approach. Build Cities 2(1):150–165
Venter ZS, Barton DN, Gundersen V, Figari H, Nowell M (2020). Urban nature in a time of crisis: recreational use of green space increases during the COVID-19 outbreak in Oslo, Norway. Environ Res Lett 15 (10):104075
von Seidlein L, Alabaster G, Deen J, Knudsen J (2020) Crowding has consequences: prevention and management of COVID-19 in informal urban settlements. Build Environ, p 107472
Wandersman A, Nation M (1998) Urban neighborhoods and mental health: psychological contributions to understanding toxicity, resilience, and interventions. Am Psychol 53(6):647
Wheaton WC (1998) Land use and density in cities with congestion. J Urban Econ 43(2):258–272
Wilkinson A (2020) Local response in health emergencies: key considerations for addressing the COVID-19 pandemic in informal urban settlements. Environ Urban 32(2):503–522
You H, Wu X, Guo X (2020) Distribution of COVID-19 morbidity rate in association with social and economic factors in Wuhan, China: implications for urban development. Int J Environ Res Public Health 17 (10):3417

4 Resilient Urban Form: There is No Myth in Enhancing Resilience Through Urban Morphologies

In the hardest times, unity is difficult but essential.
—Ali Cheshmehzangi, 2020

4.1 A General Overview of Resilient Urban Form and Morphologies

A major part of the physical dimension to discuss the health–sustainability–peace (HSP) nexus is related to urban form and morphological considerations in cities and communities. The relationship between urban form and resilience is widely recognised as part of achieving sustainability in urbanism (Sharifi and Yamagata 2018), helping us to achieve spatial morphology and resilient urban systems (Marcus and Colding 2014). Also, in relation to the pandemic events, resilient urban form has been identified as a suitable physical form of the city that helps to "*prevent infectious disease transmissio*n" (Lak et al. 2020). The idea is not necessarily news as it has been covered widely across resilience or resiliency research fields. To highlight a few in the literature, the topic of "resilient urban form" has been widely explored in areas such as urban cooling strategies (Cheshmehzangi and Butters 2017), urban energy (Yang 2015), climate change (Hamin and Gurran 2009), and climate-resilient approaches (Raven 2011). A recent macro-scale analysis by Sharifi (2019) indicates the relationship between form and resilience, specific urban form elements related to the resilience of cities, landscape connectivity, and towards key areas of adaptability, diversity, and redundancy.

In their study on introducing antifragile urban form, Sartorio et al. (2021) argue how such form could underpin societies' resilience. In resilience-pandemic research, the urban form plays a major part, exposing the vulnerability and resiliency of urban systems, urban form, and planning (Banai 2020; Cheshmehzangi 2020a). As a response, the need for transformative governance is essential, but at the physical dimension, we could explore opportunities of what we mean by resilient urban form. Some recent studies suggest ideas towards shaping a new resilient urban planning approach (Raj et al. 2021), arguing in favour of redesigning the urban form (Cheshmehzangi 2021; Yamagata and Yoshida 2020), and towards resilient societies (Douglas et al. 2020). Therefore, urban morphologies are part of a larger network of the physical dimension of cities and the networks of socio-economic factors, transportation, and other related urban systems (Wang et al. 2017).

As a major understanding of the physical dimension in achieving the HSP nexus, we note the relevance of the topic regarding the human aspects of urban form, which is narrated from a man–environment approach in urbanism (Rapoport 2016).

A. Cheshmehzangi, *Urban Health, Sustainability, and Peace in the Day the World Stopped*, Sustainable Development Goals Series, https://doi.org/10.1007/978-981-16-4888-5_4

This ideology is partly based on the earlier work on the anthropology of urban form (Rykwert 1988), which is also discussed in the urban planning and peacebuilding research (Bollens 2006). In an example of remaking Chinese urban form (Lu 2006), we also see some similarities in identifying spatial and physical attributes related to redefining the urban built environments. In his book on "Uneasy Peace", Sharkey (2018) explores the relationship between urban life, crime and violence issues, and maintaining peace in cities. This is closely related to key areas in building an inclusive peace through the consideration of socio-spatial processes (Koopman 2020), especially from the perspectives of equality, accessibility to resources, and reflecting on spatialised social inequalities. The socio-spatial processes are certainly embedded in the physicality of the urban form, allowing for the resiliency of networks and communities. In this regard, we argue in favour of resilient urban forms, also strengthening the peacebuilding opportunities through the physical–spatial network of cities and communities.

4.2 Reflecting on How Cities Reacted During the Pandemic

Cities are rich with urban landscapes that could be the setting for peacebuilding strategies. The COVID-19 pandemic opened up many opportunities for us to reinterpret conventional urban planning and design strategies, particularly from the physical dimension. By considering the urban futures and enhancing peace (or "non-violence") in communities, we are able to find pathways for future landscapes (Hutchinson and Herborn 2012) that could become the backbone of considering better urban forms and morphologies. In some of our findings, we have noted a strong relationship between urban morphologies and resilience, especially during pandemic events. The spatiality of the urban form plays a significant part in how activities and social networks reacted during the pandemic. Partly related to the previous chapter on urban density matters, we note the relationships that exist between urban geometry, socio-demographic characteristics, and spatial qualities of the urban form (Kwok et al. 2021). In this regard, cities' urban morphology determines primary built environment factors such as building shapes, road and street networks, economic characteristics (ibid), and other key aspects of urban layout patterns.

For instance, in poorer communities, we identify a different type of urban form that determines specific spatial qualities and inequalities. This is particularly in living or residential areas, where significant disparities are identified to impact health-related matters for the residents (Cheshmehzangi 2020b). Issues related to natural environments and healthy communities are those that are highlighted well in research (de Vries et al. 2003; Bell et al. 2014; Navarrete-Hernandez and Laffan 2019; Xue et al. 2019; Cheshmehzangi 2020b). For example, during the COVID-19 pandemic, many research findings showed the positive impact of urban form with green spaces on people's health and well-being (Poortinga et al. 2021). Some studies also correlated such health-oriented matters to other factors, showing benefits for mental health (Pouso et al. 2021) as well as physical health and better emotions/feelings (Ugolini et al. 2021).

One of our key findings of earlier studies related to urban form relates to the actual typology of the urban form, especially discussing more resilient morphologies (Cheshmehzangi et al. 2021). For instance, despite some of the negativities of the gated communities (Cheshmehzangi 2018), we found out positive aspects of such form in maintaining the resilience of communities (Mangi et al. 2021). Therefore, urban form plays a major part in sustaining health and peace in cities and communities, especially in compact living environments that are interlinked with other networks and urban systems. In many examples, the urban form determines opportunities for control and management of communities, networks of mobility between them, and the connectivity that goes beyond just the physicality of the place. In the cases of compact urban environments, the form is indeed very critical in containing epidemic spread. For instance, urban forms with particular spatial characteristics suggest open space layout, singular or multiple entries and exit points (Cheshmehzangi 2020c), places of social interaction, availability of and accessibility to green spaces, and some other design and planning factors.

During the COVID-19 pandemic, we noted that some urban morphologies are more resilient than others (Mangi et al. 2021). In fact, we also highlighted the case is strongly correlated with two other factors of urban land use and density, which are covered in the previous chapter. Moreover, from the physical dimension, the consideration of HSP nexus could be beneficial to enhance the well-being of communities. While closed and open residential communities differ in terms of their spatiality, we can determine better resilience through morphologies. The same principles exist in adaptive urbanism (Fields and Renne 2021), where we could be related to other concepts such as stability and sustainability of urban forms through the role of place (Mai 2020). More importantly, urban resilience is not just an urban management perspective as defined by McGill (2020), but rather a complementary tool to city management (Cheshmehzangi 2020a). In this regard, resilience through urban form could be an instrumental tool (Brunetta et al. 2021), which could help to improve the urban fabric, spatial layout, and network of spaces that exist within and in between urban communities (and living environments).

4.3 Addressing the Flaws

The prescriptive approach to suggest specific urban morphologies could be a treacherous act, which may become a misguidance to planners and designers or become a wrong tool in the hands of the government officials and developers. In the arguments of preparedness through urban resilience (Cheshmehzangi 2020d), straightforward suggestions are given to ensure resilience and health are enhanced within the fabric of the cities. As mentioned by Moss (2020, p. 212), residents often "*cope with physical, economic, and political disruptions*", through which we can see multiple vulnerabilities and a range of challenges in this complexity. Through the physical realm, we are able to respond to some of these complexities, resolving the way cities are shaped and continuously reshaped.

As a general approach, the upgrade of urban form is essential to respond to unhealthy urban conditions. This approach is mainly effective for poorer communities or those areas that face inequalities in terms of physical infrastructures, necessary amenities, etc. To enhance peace and achieve healthy living environments, one ought to consider the position of robust urban management, which could be achieved not necessarily just through closed urban enclaves, but also through new paradigms of hybrid urban patterns. By reflecting on the spatial qualities of gated communities, we could learn how to make vibrant communities while keeping the residential areas safe and secure. This does not have to happen through the boundaries of a gated community. In a good urban design, we are able to utilise and combine morphological, socio-spatial, and temporal dimensions to ensure urban communities are both vibrant and healthy communities in terms of their resilience. A major application of such an approach would be on slum upgrading programmes or projects that aim to enhance the city's health and societal well-being, particularly through the physical upgrades in cities. In some contexts, such upgrades are phrased as modernisation processes, which may give a wrong perception to the general public. Instead, we need to see how existing urban forms perform in their specific locales, what do they offer in terms of spatial qualities, socio-cultural interactions, and other related configurations, and consider methods of improvement, which does not mean significant changes to the urban fabric.

In general, urban resilience and urban sustainability are interrelated. For us to maintain a decent HSP nexus, we could also utilise smarter technologies and smart solutions, which are discussed in Part 6 of the study. In order to define a good urban morphology, we ought to consider the context-specific factors that may be only specific to certain groups, cities, or regions/countries. However, it is essential to also consider paradigm shifts that could help to enhance the resilience of cities and communities (as well as neighbourhoods, urban blocks, etc.) and develop/propose new urban morphologies.

4.4 A Summary

Cities are shaped by their urban landscapes and the physicality of the urban environments that ultimately become the network of places and living environments. Some urban forms tend to become more prevalent in specific contexts, and some become trendy global models that may be wrongly duplicated in different contexts. Our role here is not to create a set of doctrines for specific urban forms but to consider the importance of urban morphologies in achieving the HSP nexus. In this regard, we aim to interpret what becomes the health determinants of suitable urban morphology in cities where sustainable urban forms are non-existent. Such interpretation needs to involve both dimensions of spatio-temporal and morphological, helping to inform decision-making processes for urban planning and design.

As an important physical attribute, urban form is closely correlated with public health matters (Vojnovic et al. 2006; Bejleri et al. 2016; Cheshmehzangi and Butters 2017; Butters et al. 2020). This correlation is studied extensively regarding inclusive design approaches, integrated planning strategies, and the inclusion of greenery and natural environments in compact living environments. When it comes to pandemic events, like the COVID-19 pandemic, social determinants of health and health equity matters are as important as linking the urban form with quality living environments. In this regard, we have to consider urban health adaptation strategies that could inform design and planning processes, new urban layouts, and better physical infrastructures. We follow some of these in the following chapter, in which we stress the role of critical infrastructures in maintaining a reasonable HSP nexus in cities and communities.

References

Banai R (2020) Pandemic and the planning of resilient cities and regions. Cities 106:102929

Bejleri I, Steiner R, Yoon S, Neff D, Harman J (2016) A93—does health disparity differ by urban form type? A comparative analysis of urban, suburban, and rural areas. J Transp Health 3(2):S60--S61

Bell SL, Phoenix C, Lovell R, Wheeler BW (2014) Green space, health and wellbeing: making space for individual agency. Health Place 30:287–292

Bollens SA (2006) Urban planning and peace building. Prog Plan 66(2):67–139

Brunetta G, Faggian A, Caldarice O (2021) Bridging the gap: the measure of urban resilience. Sustainability 13 (3):1113. https://doi.org/10.3390/su13031113

Butters C, Cheshmehzangi A, Sassi P (2020) Cities, energy and climate: seven reasons to question the dense high-rise city. J Green Build 15(3):197–214

Cheshmehzangi A, Butters C (eds) (2017) Designing cooler cities: energy, cooling and urban form—the Asian perspective. Palgrave Macmillan, Singapore

Cheshmehzangi A (2018) The changing urban landscape of Chinese cities: positive and negative impacts of urban design controls on contemporary urban housing. Sustainability 10(8):2839. https://doi.org/10.3390/su10082839

Cheshmehzangi A (2020) Housing and health evaluation related to general comfort and indoor thermal comfort satisfaction during the COVID-19 lockdown. J Hum Behav Soc Environ 31(1–4):184–209

Cheshmehzangi A (2020) The city in need: urban resilience and city management in disruptive disease outbreak events. Springer, Singapore

Cheshmehzangi A, Botti G, Mangi E, Dawodu A, Sedrez M (2021).Urban form in China: entering an era of more ‘Gated Communities’ for resilient urbanism. Arch J Sri Lankan Institute of Architects, pp 234–239

Cheshmehzangi A (2020c) 10 Adaptive measures for public places to face the COVID 19 pandemic outbreak. City Soc. Special Issue, Available from: https://doi.org/10.1111/ciso.12282

Cheshmehzangi A (2020d) Preparedness through urban resilience. In: Cheshmehzangi A (ed) The city in need: urban resilience and city management in disruptive disease outbreak events. Springer, Singapore. https://doi.org/10.1007/978-981-15-5487-2_3

Cheshmehzangi A (2021) Revisiting the built environment: 10 potential development changes and paradigm shifts due to COVID-19. J Urban Manage. Available from: https://www.sciencedirect.com/science/article/pii/S2226585621000054

de Vries S, Verheij RA, Groenewegen PP, Spreeuwenberg P (2003) Natural environments—healthy environments? An exploratory analysis of the relationship between greenspace and health. Environ Plan A Econ Space 35(10):1717–1731

Douglas I, Champion M, Clancy J, Haley D, de Souza ML, Morrison K, Scott A, Scott R, Stark M, Tryjanowski P, Webb T (2020) The COVID-19 pandemic: local to global implications as perceived by urban ecologists. Socio-Ecol Pract Res 2(3):217–228

Fields B, Renne JL (2021) From urban resilience to street resilience. In: Fields B, Renne JL (eds) Adaptation urbanism and resilient communities: transforming streets to address climate change. Routledge, New York, 23 pages. https://doi.org/10.4324/9780429026805

Hamin EM, Gurran N (2009) Urban form and climate change: balancing adaptation and mitigation in the US and Australia. Habitat Int 33(3):238–245

Hutchinson FP, Herborn PJ (2012) Landscapes for Peace: a case study of active learning about urban environments and the future. Futures 44(1):24–35

Koopman S (2020) Building an inclusive peace is an uneven socio-spatial process: Colombia's differential approach. Polit Geogr 83:102252

Kwok CYT, Wong MS, Chan KL, Kwan MP, Nichol JE, Liu CH, Wong JYH, Wai AKC, Chan LWC, Xu Y, Li H, Huang J, Kan Z (2021). Spatial analysis of the impact of urban geometry and socio-demographic characteristics on COVID-19, a study in Hong Kong. Sci Total Environ 764:144455

Lak A, Asl SS, Maher A (2020) Resilient urban form to pandemics: lessons from COVID-19. Med J Islam Repub Iran 34:71

Lu D (2006) Remaking Chinese urban form: modernity, scarcity and space, 1949–2005. Routledge, New York and London

Mai X (2020) Spotlight on… urban resilience: Is it old wine in a new bottle? Geography 105(3):161–166

Mangi E, Cheshmehzangi A, Botti G (2021) COVID-19, gated communities, and urban resilience: a comparative outlook between China and Colombia. In: Montoya MA, Rehner J, Kristikj A, Lemus-Delgado D (eds) COVID-19 and cities: experiences, responses, and uncertainties. Springer, Cham, The Urban Book Series, in press

Marcus L, Colding J (2014) Toward an integrated theory of spatial morphology and resilient urban systems. Ecol Soc 19(4)

McGill R (2020) Urban resilience—an urban management perspective. J Urban Manage 9(3):372–381

Moss T (2020) Urban resilience has a history—and a future. In: Brantz D, Sharma A (eds) Urban resilience in a global context: actors, narratives, and temporalities. Transcript Publishing, New Rockford, pp 209–215

Navarrete-Hernandez P, Laffan K (2019) A greener urban environment: Designing green infrastructure interventions to promote citizens' subjective wellbeing. Landsc Urban Plan 191(6):103618. Article 103618

Poortinga W, Bird N, Hallingberg B, Phillips R, Williams D (2021) The role of perceived public and private green space in subjective health and wellbeing during and after the first peak of the COVID-19 outbreak. Landsc Urban Plan 211:104092

Pouso S, Borja A, Fleming LE, Gómez-Baggethun E, White MP, Uyarra MC (2021) Contact with blue-green spaces during the COVID-19 pandemic lockdown beneficial for mental health. Sci Total Environ 756:143984

Raj SA, Angella EJ, Pooja C (2021) Impact of Covid-19 in shaping new resilient urban planning approach. In: IOP conference series: materials science and engineering, vol 1114, no 1. IOP Publishing, p 012040

Rapoport A (2016) Human aspects of urban form: towards a man—environment approach to urban form and design. Elsevier, Amsterdam

Raven J (2011) Cooling the public realm: climate-resilient urban design. In: Resilient cities. Springer, Dordrecht, pp 451–463

Rykwert J (1988) The idea of a town: the anthropology of urban form in Rome, Italy and the ancient world. MIT Press, Cambridge

Sartorio FS, Aelbrecht P, Kamalipour H, Frank A (2021) Towards an antifragile urban form: a research agenda for advancing resilience in the built environment. Urban Des Int, pp 1–24

Sharifi A (2019) Resilient urban forms: a macro-scale analysis. Cities 85:1–14

Sharifi A, Yamagata Y (2018) Resilient urban form: a conceptual framework. In: Resilience-oriented urban planning. Springer, Cham, pp 167–179

Sharkey P (2018) Uneasy peace: the great crime decline, the renewal of city life, and the next war on violence. WW Norton & Company, New York

Ugolini F, Massetti L, Pearlmutter D, Sanesi G (2021). Usage of urban green space and related feelings of deprivation during the COVID-19 lockdown: lessons learned from an Italian case study. Land Use Policy 105:105437

Vojnovic I, Jackson-Elmoore C, Holtrop J, Bruch S (2006) The renewed interest in urban form and public health: Promoting increased physical activity in Michigan. Cities 23(1):1–17

Wang S, Liu X, Zhou C, Hu J, Ou J (2017) Examining the impacts of socioeconomic factors, urban form, and transportation networks on CO_2 emissions in China's megacities. Appl Energy 185:189–200

Xue F, Lau SS-Y, Gou Z, Song Y, Jiang B (2019) Incorporating biophilia into green building rating tools for promoting health and wellbeing. Environ Impact Assess Rev 76:98–112

Yamagata Y, Yoshida T (2020) A "Smart Lifestyle" for the re-design of the "After Corona" urban form. Environ Plann B Urban Anal City Sci 47(7):1146–1148

Yang PPJ (2015) Energy resilient urban form: a design perspective. Energy Procedia 75:2922–2927

Critical Infrastructures and Safeguarding the City's Sustainability and Peace

5

By paying too much attention to generalities, we may simply neglect the details.

—Ali Cheshmehzangi, 2018

5.1 A General Overview of Critical Infrastructures in Cities

The Cybersecurity and Infrastructure Security Agency (CISA), based in the USA, was probably the first agency that identified critical infrastructures (CIs) during the COVID-19 pandemic (CISA, 2020). In their work, they identified 16 critical infrastructure sectors, which are based on the Presidential Policy Directive 21 (PPD-21), indicating that their "*assets, systems, and networks, whether physical or virtual, are considered so vital to the United States that their incapacitation or destruction would have a debilitating effect on security, national economic security, national public health or safety, or any combination thereof*" (ibid). The 16 critical infrastructure sectors include (1) chemical sector, (2) commercial facilities sector, (3) communications sector, (4) critical manufacturing sector, (5) dams sector, (6) defence industrial base sector, (7) energy sector, (8) emergency services sector, (9) financial services sector, (10) food and agriculture sector, (11) government facilities sector, (12) healthcare and public health sector, (13) information technology sector, (14) nuclear reactors, materials, and waste sector, (15) transportation system sector, and (16) water and wastewater systems sector. In addition, they promote essential guidelines for the workers of these sectors and highlight the key industry and government responses to the essential critical infrastructure workforce, which are the backbone of city operations (Cheshmehzangi 2020a), especially during calamitous and emergency situations like a pandemic.

One correlation that is discussed widely is between CIs and urban health, such as for dealing with issues of disaster management (O'Sullivan et al. 2013), considering health impact assessment (Browne and Lowe 2021), infrastructure safety (Valero et al. 2021), and emergency and business continuity (Galbusera et al. 2021). As the scholarly research suggests, we continuously seek to strengthen organisational resilience in our critical infrastructure systems (Rehak 2020), which is incredibly important in dealing with a variety of threats, from climatic and natural disasters to health emergencies, terrorism, etc. While many studies cover the correlation well, there is little literature on how we can foster peace through the capacity of our critical infrastructures. Only very few studies look into methods of rethinking development by considering peace engineering (Kleba and Reina-

A. Cheshmehzangi, *Urban Health, Sustainability, and Peace in the Day the World Stopped*, Sustainable Development Goals Series, https://doi.org/10.1007/978-981-16-4888-5_5

Rozo 2021), considering the sustainable development goals (Yarnall et al. 2021), or specific innovation systems (Ahn and Yi 2021), which are considerably different to what we really mean by achieving urban health in practice. As Galbusera et al. (2021) define them well, CIs are almost like our societies' nerve/nervous system, which should indeed respond to people's needs and issues. They need to be able to provide critical supplies (Emmanuel et al. 2020; Ranney et al. 2020) and ensure the residents receive their essential services and provisions (Cheshmehzangi 2020a). In this regard, we argue in favour of the broadening perspectives (Galbusera et al. 2021), which would lead to identifying critical services (Bahmanyar et al. 2020), risk management, and responsive behaviour, to control and contain the emergency situations like the COVID-19 pandemic. In all conditions, the continuity of businesses for essential services and critical infrastructures (Koonin 2020) and city operations (Cheshmehzangi 2020a) are crucial to sustaining health and peace in cities and communities.

As identified during the COVID-19 pandemic, an important part of critical infrastructure is the availability of and accessibility to specific services, such as healthcare services, safety and security services, and social and public services. The combined effect of shortages (in supplies/resources, workforces, etc.) and deficiencies could lead to prolonged adversities that may end up in much harder disruptive situations. While most of the earlier literature related to critical infrastructure and peace are based on cyber-peace and security issues (Condron 2006; Collier and Lakoff 2008; Lundborg and Vaughan-Willians 2011; Shackelford 2011; Sheldon 2011; Preciado 2012; Atlas 2013; Quigley et al. 2015), we see growing research on nationwide systems and protecting them in specific occasions (Lewis 2019). It has been proven that we could build security through critical infrastructure and achieve peace (Zeideraman 2020). In doing so, we are able to consider sustainable peace through healthy development (Igbuzor 2011), which will be part of the city's/community's physical dimension.

5.2 Reflecting on How Cities Reacted During the Pandemic

During the COVID-19 pandemic, many people and governments realised the importance of critical infrastructure in containing the disease. In fact, having concrete critical infrastructures would enable us to have a much stronger preparedness plan against pandemic events (Cheshmehzangi 2020b). Some studies looked at the ecological and environmental aspects of cities, such as the importance of forests as critical infrastructure (Derks et al. 2020) and their role in maintaining healthy mobility and movement. Other studies looked at health-related workspace issues (Groenewold et al. 2020), realising that a large body of our population work in essential services and critical infrastructure (Mahase 2020). The vulnerability of specific essential critical infrastructures was highlighted in various cases, such as in healthcare homes (Boal et al. 2021), food and agricultural sectors (Walters et al. 2020), and health infrastructures (Barabadi et al. 2020). More important than all were the socio-economic determinants of the pandemic (Stojkoski et al. 2020), which enabled us to widen our horizons on the importance of critical infrastructure in emergencies.

As many cities started to receive many infected patients, their capacity of healthcare units and services could no longer fulfil the needs of society. Many countries adopted other methods, such as home treatments, temporary treatment centres, and mobile services that could help the overflowing healthcare services. Apart from the health care, other CIs also faced difficulties as well. Some will be discussed in the economic part of the book, while the others are more relevant to the physicality of the urban environments/settings. Cities are known to be places of innovation. Still, during the pandemic, we realised the many flaws in our everyday operations and preparedness that become a necessity against emergencies. Many cities and regions utilised the capacity of task force groups to better respond to sector-based situations, mainly aiming to support essential and critical services, such as healthcare units, hospitals, and clinics.

If we have to evaluate fairly, we would argue that many cities failed in managing the COVID-19 pandemic. Much of the failure cases were due to a lack of critical infrastructures, lack of preparedness, or lack of support that should be in place in such events. Cities with prior experiences and immediate response and adaptive measures could respond to the situation more effectively. On the contrary, many cities left their critical infrastructure with ineffective operations. Some led to a lack of support protocol, processes, legal, and regulatory protections (Maves et al. 2020), which are essential to safeguard the city and its operations. Some recent studies suggest high-level improvements or upgrades, such as the need for a national health information technology infrastructure (Sittig and Singh 2020), which would possibly help optimise the overall management of the healthcare systems. Other examples looked at revisiting the physicality of the built environments (Cheshmehzangi 2021), hoping for substantial change and/or transition to enhance the well-being and health of communities through better urban infrastructures and systems.

One of the areas that require immediate attention is the protection of the CIs in future events (Kortuem et al. 2020). The fact that cities had to repurpose and reshape their hospitals during the COVID-19 (Her 2020) highlights the importance of adaptive planning or adaptiveness in general. The example of increasing telemedicine (Chauhan et al. 2020) indicates opportunities that require transformative decision-making, enabling us to better protect the CIs and workforces of specific sectors. Several years ago, Rehak et al. (2016) studied the synergistic issues in our critical infrastructure system, identifying many failures and issues that undermine the role of CIs in protecting the city and representing an interconnected system. Their approach to critical infrastructure systems at the national level is holistic, suggesting three levels: system level, sector level, and element level (ibid). This hierarchical arrangement enables us to verify the complexities of technical infrastructure and socio-economic infrastructure (ibid) embedded in critical infrastructures' operations. Their findings highlight the correlation that exists strongly between the intensity of cause, failure, and impact in a critical infrastructure system (ibid), which also is a good representation of what happened during the COVID-19 pandemic. Under the CIs, we see the complexity of multi-spatial situations, often through physical connections and other networks; thus, they become even more important to highlight influences, dependencies, and independencies of the urban systems. The detailed findings of their study (ibid) are extremely useful for cities that need to create action plans to support their CI system. Hence, they propose five aspects to evaluate and create the character of impacts in a CI system, including (1) area of the action, (2) structure of the action, (3) intensity of the action, (4) duration, and (5) effect of the action (ibid). For health emergencies, like the COVID-19 pandemic, this model becomes very useful as we can evaluate sectors and crossovers between them to ensure synergies are developed or maintained and effective action plans are suggested.

5.3 Addressing the Flaws

Undoubtedly, our CIs are the backbone of city operations and day-to-day activities that health and peace are sabotaged without them. By considering the system preparedness and strengthening the synergies between them, we can enhance public health through the physicality of the built environments (Cheshmehzangi 2020c), and more importantly, through the capacity of critical infrastructures. For long, it is evident that more advanced systems could support the operations in cities, such as the use of telecommunications infrastructure in stabilising economic development (Roller and Waverman 2001) or finding methods that could bridge the gaps between people, perspectives, and planning

(Cheshmehzangi and Dawodu 2019). During the COVID-19 pandemic, it became even more evident that technology-based interventions (Su et al. 2020) could better manage the conventional practices and enhance communities' well-being and health.

By addressing the flaws of the CIs, we ought to evaluate the quantity and quality of what we have and how they operate. It is then important to invest in those failed CI systems, create better synergies in their operations, and refine them as the backbone of peace-making in city operations. From energy systems (Cheshmehzangi 2020d) to waste management systems (Dawodu et al. 2021), we ought to be reflective and progressive in terms of not just experimentations but interventions that could change the current so-called normal operations. As an immediate response, we have to enhance the main features of our emergency communication (Khayat et al. 2019), which should assure peace through relevant procedures. In other responses, we must invest in protecting the CIs (Ellis et al. 1997), something that for long has become a challenge for many cities and regions around the globe. As we move forward, we need to enhance our capacity for risk management, ensuring that our systems are integrated but protected (Moteff 2005). To address this more effectively, we need to protect our assets and transmissions (ibid) and provide adequate support to upgrade the operational mechanisms.

In the following steps, should we come out of the COVID-19 pandemic any time soon, we must consider integrated systems that are safe, effective, and responsive. The use of ICT-mediated systems and smart technologies (see Part 6) is suitable examples of enhancing our security management, develop a better information network, and protect the daily operations of our cities and communities. In the pandemic, critical infrastructure receives the highest level of impact, showing that we generally rely on them (Itzwerth et al. 2006) to maintain peace and urban health.

5.4 A Summary

During the COVID-19 pandemic, several CIs appeared to be more important than the others. Those specific CIs related to public services and provisions are generally the ones that are more needed in cities with higher risk levels. Specific CIs related to safety and security services, healthcare services, and transportation services are of primary concern, especially in places where poverty is widespread. In cities of unplanned structures or with lower quality planning frameworks, we note significant issues with CIs such as transportation services and healthcare services. In the COVID-19 pandemic, it becomes evident that CIs are essential to every single city's operations.

We further express that a city is a place for experimentation and a place for innovation and intervention. As we close the first part of the book, we highlight the many flaws of our physical built environments, some that could lead to chaotic situations and societal unrest. The city's health depends on its quality urban environments, spaces, and many realms that relate to our daily operations and activities. If they are not safeguarded, we may put more pressure on the sustainability and peace of the city. As societies faced and continue to face the COVID-19 pandemic, we explore many examples of spatial and physical flaws that could only be addressed if better urban planning/design and policies are in place. Some issues are contextual indeed (Arakpogun et al. 2020), but some are very much related to issues that undermine the quality and effectiveness of our CIs. We also note the importance of environmental aspects in developing or maintaining a good physical environment, some that may relate to overarching aspects of sustainable development (Marco et al. 2020) and risk reduction in our living environments. Lastly, we note that the relationship between physical and environmental dimensions are very strong; and hence, we would delve into some of the environmental aspects in the next part of the book.

References

Ahn S-J, Yi S-K (2021) Methodological framework for analyzing peace engineering: focusing on Kaesong industrial complex and North Korean innovators in South Korea. Technol Forecast Soc Change 163:120464

Arakpogun E, Sahn ZE, Prime KS, Gerli P, Olan F (2020) Africa's resilience in the face of COVID-19 pandemic: let's talk about It!. 10 pages, Available at SSRN 3640311. https://papers.ssrn.com/sol3/papers.cfm?abstract_id=3640311

Atlas RI (2013) 21st century security and CPTED: designing for critical infrastructure protection and crime prevention. CRC Press, Boca Raton

Bahmanyar A, Estebsari A, Ernst D (2020) The impact of different COVID-19 containment measures on electricity consumption in Europe. Energ Res Soc Scie 68:101683

Barabadi A, Ghiasi MH, Qarahasanlou AN, Mottahedi A (2020) A Holistic View of Health Infrastructure Resilience before and after COVID-19. Arch Bone Joint Surg 8(Suppl 1):262

Boal WL, Li, J, Silver SR (2021) Health care access among essential critical infrastructure workers, 31 States, 2017–2018. Public Health Rep 0033354921996688

Browne GR, Lowe M (2021) Liveability as determinant of health: testing a new approach for health impact assessment of major infrastructure. Environ Impact Assess Rev 87:106546

Chauhan V, Galwankar S, Arquilla B, Garg M, Di Somma S, El-Menyar A, Stawicki SP et al (2020) Novel coronavirus (COVID-19): leveraging telemedicine to optimize care while minimizing exposures and viral transmission. J Emergencies, Trauma Shock 13 (1):20

Cheshmehzangi A (2020a) The city in need: urban resilience and city management in disruptive disease outbreak events. Springer, Singapore

Cheshmehzangi A (2020b) Preparedness through urban resilience. In: Cheshmehzangi A (ed) The city in need: urban resilience and city management in disruptive disease outbreak events (Chapter 3). Springer, Singapore. https://doi.org/10.1007/978-981-15-5487-2_3

Cheshmehzangi A (2020c) Reflection on early lessons for urban resilience and public health enhancement during the COVID-19. Health 12(10):1390

Cheshmehzangi A (2020d) COVID-19 and household energy implications: what are the main impacts on energy use? Heliyon 6(10):e05202

Cheshmehzangi, A. (2021). Revisiting the built environment: 10 potential development changes and paradigm shifts due to COVID-19. J Urban Manage. https://www.sciencedirect.com/science/article/pii/S2226585621000054

Cheshmehzangi A, Dawodu A (2019) Case study reviews: people, perspective and planning. In: Cheshmehzangi A, Dawodu (eds) Sustainable urban development in the age of climate change: people: the cure or curse. Palgrave Macmillan, Singapore, pp 69–131

CISA (Cybersecurity & Infrastructure Security Agency) (2020) Identifying critical infrastructure during COVID-19. https://www.cisa.gov/identifying-critical-infrastructure-during-covid-19

Collier S, Lakoff A (2008) The vulnerability of vital systems: How 'critical infrastructure' became a security problem. The politics of securing the homeland: critical infrastructure, risk and securitisation, pp 40–62

Condron SM (2006) Getting it right: protecting American critical infrastructure in cyberspace. Harvard J Law Techol 20(2):403–422

Dawodu A, Oladejo J, Tsiga Z, Kanengoni T, Cheshmehzangi A (2021) Underutilization of waste as a resource: bottom-up approach to waste management and its energy implications in Lagos, Nigeria. Intelligent Buildings International, pp 1–22

Derks J, Giessen L, Winkel G (2020) COVID-19-induced visitor boom reveals the importance of forests as critical infrastructure. Forest Policy Econ 118:102253

Di Marco M, Baker ML, Daszak P, De Barro P, Eskew EA, Godde CM, Harwood TD, Herrero M, Hoskins AJ, Johnson E, Karesh WB (2020) Opinion: sustainable development must account for pandemic risk. Proc Natl Acad Sci 117(8):3888–3892

Ellis J, Fisher D, Longstaff T, Pesante L, Pethia, R (1997) Report to the president's commission on critical infrastructure protection. Carnegie-Mellon University, Pittsburgh PA Software Engineering Institute. https://apps.dtic.mil/sti/citations/ADA324232

Emanuel EJ, Persad G, Upshur R, Thome B, Parker M, Glickman A, Zhang C, Boyle C, Smith M, Phillips JP (2020) Fair allocation of scarce medical resources in the time of Covid-19. N Engl J Med 382:2049–2055

Galbusera L, Cardarilli M, Giannopoulos G (2021) The ERNCIP survey on COVID-19: emergency & business continuity for fostering resilience in critical infrastructures. Saf Sci 105161

Groenewold MR, Burrer SL, Ahmed F, Uzicanin A, Free H, Luckhaupt SE (2020) Increases in health-related workplace absenteeism among workers in essential critical infrastructure occupations during the covid-19 pandemic—United States, March–April 2020. Morb Mortal Wkly Rep 69(27):853

Her M (2020) Repurposing and reshaping of hospitals during the COVID-19 outbreak in South Korea. One Health 10:100137

Igbuzor O (2011) Peace and security education: a critical factor for sustainable peace and national development. Int J Peace Dev Stud 2(1):1–7

Itzwerth RL, MacIntyre CR, Shah S, Plant AJ (2006) Pandemic influenza and critical infrastructure dependencies: possible impact on hospitals. Med J Aust 185 (S10):S70–S72

Khayat G, Mavromoustakis CX, Batalla JM, Mastorakis G, Maalouf H, Mukherjee M (September 2019). Tuning the uplink success probability in damaged critical infrastructure for VANETs. In:

2019 IEEE 24th international workshop on computer aided modeling and design of communication links and networks (CAMAD). IEEE, pp 1–6

Kleba JB, Reina-Rozo JD (2021) Fostering peace engineering and rethinking development: a Latin American view. Technol Forecast Soc Change 167:120711

Koonin LM (2020) Novel coronavirus disease (COVID-19) outbreak: now is the time to refresh pandemic plans. J Bus Contin Emer Plan 13:1–15

Kortuem SO, Becker D, Ott HJ, Schlaudt HP (2020) The role of the emergency department in protecting the hospital as a critical infrastructure in the corona pandemic strategies and experiences of a rural sole acute-care clinic. medRxiv

Lewis TG (2019) Critical infrastructure protection in homeland security: defending a networked nation. Wiley, New York

Lundborg T, Vaughan-Williams N (2011) Resilience, critical infrastructure, and molecular security: The excess of "life" in biopolitics. Int Political Sociol 5 (4):367–383

Mahase E (2020) Covid-19: outbreak could last until spring 2021 and see 7.9 million hospitalised in the UK. https://www.bmj.com/content/368/bmj.m1071/

Maves RC, Downar J, Dichter JR, Hick JL, Devereaux A, Geiling JA, Kissoon N, Hupert N, Niven AS, King MA, Rubinson LL (2020) Triage of scarce critical care resources in COVID-19: an implementation guide for regional allocation: an expert panel report of the task force for mass critical care and the american college of chest physicians. Chest

Moteff J (February 2005) Risk management and critical infrastructure protection: assessing, integrating, and managing threats, vulnerabilities and consequences. Library of Congress Washington DC Congressional research service. https://apps.dtic.mil/sti/citations/ADA454038

O'Sullivan TL, Kuziemsky CE, Toal-Sullivan D, Corneil W (2013) Unraveling the complexities of disaster management: a framework for critical social infrastructure to promote population health and resilience. Soc Sci Med 93:238–246

Preciado M (2012) If you wish cyber peace, prepare for cyber war: the need for the federal government to protect critical infrastructure from cyber warfare. J Law Cyber Warfare 1(1):99–154

Quigley K, Burns C, Stallard K (2015) 'Cyber Gurus': a rhetorical analysis of the language of cybersecurity specialists and the implications for security policy and critical infrastructure protection. Gov Inf Q 32(2):108–117

Ranney ML, Griffeth V, Jha AK (2020) Critical supply shortages—the need for ventilators and personal protective equipment during the Covid-19 pandemic. The New England J Med 382(18):e41

Rehak D (2020) Assessing and strengthening organisational resilience in a critical infrastructure system: case study of the Slovak Republic. Saf Sci 123:104573

Rehak D, Markuci J, Hromada M, Barcova K (2016) Quantitative evaluation of the synergistic effects of failures in a critical infrastructure system. Int J Crit Infrastruct Prot 14:3–17

Roller LH, Waverman L (2001) Telecommunications infrastructure and economic development: a simultaneous approach. Am Econ Rev 91(4):909–923

Shackelford SJ (2011) In search of cyber peace: a response to the cybersecurity act of 2012. Stanford Law Rev Online 64:106–110

Sheldon JB (2011) Deciphering cyberpower: strategic purpose in peace and war. Strateg Stud Q 5(2):95–112

Sittig DF, Singh H (2020) COVID-19 and the need for a national health information technology infrastructure. JAMA 323(23):2373–2374

Stojkoski V, Utkovski Z, Jolakoski P, Tevdovski D, Kocarev L (2020) The socio-economic determinants of the coronavirus disease (COVID-19) pandemic. arXiv preprint arXiv:2004.07947

Su Z, Meyer K, Li Y, McDonnell D, Joseph NM, Li X, Du Y, Advani S, Cheshmehzangi A, Ahmad J, da Veiga CP (2020) Technology-based interventions for nursing home residents: implications for nursing home practice amid and beyond the influence of COVID-19: a systematic review protocol

Valero M, Li F, Zhao L, Zhang C, Garrido J, Han Z (2021) Vibration sensing-based human and infrastructure safety/health monitoring: a survey. Digital Signal Process 114:103037

Walters L, Wade T, Suttles S (2020) Food and agricultural transportation challenges amid the COVID-19 Pandemic. Choices 35(3)

Yarnall K, Olson M, Santiago I, Zelier C (2021) Peace engineering as a pathway to the sustainable development goals. Technol Forecas Soc Change 168:120753

Zeiderman A (2020) Concrete peace: building security through infrastructure in Colombia. Anthropol Q 93 (3):497–528

Part II

Sustainability and Peace: "Urban Health from the Environmental Dimension"

In the second part of studying sustainability and peace, we present three primary aspects representing urban health from the environmental dimension.

The first aspect is focused on the overarching ideas of a well-functioning government, specifically for achieving peaceful living environments. This is also one of the main pillars of positive peace. In Chap. 6, we particularly explore some of the key features that define governance's role in maintaining peace and sustainability in cities and communities. The second aspect is closely linked with Chap. 5 of the book on critical infrastructures. Instead, we focus more on essential systems and operations under the environment dimension than the actual (physical) infrastructures. However, the two are closely linked as they are complementary to each other. These aspects will be highlighted in Chap. 7, through which we aim to reflect on some of the city's essential systems. The third aspect is focused on resource efficiency and distribution, which is partly related to one of the positive peace pillars. Here, we highlight the diversity of "resources", focusing more on healthy environments and resourceful societies. This important aspect is presented in Chap. 8, highlighting the role of resource efficiency and distribution in achieving urban health, sustainability, and peace.

6 Well-Functioning Government and Peaceful Living Environments

Our care to the environment is often a matter of pretend.

—Ali Cheshmehzangi, 2021

6.1 A General Overview of Well-Functioning Government for the Urban Environments

It is inevitable to consider a well-functioning government, one of the main positive peace pillars, as part of the institutional dimension of sustainability. Nonetheless, this by-default thinking may not be completely accurate. A well-functioning government, although institutionally sound, may be more relevant to enhance competencies for a successful life (Rychen and Salganik 2003), which is more relevant to the actual success of the living environments in our cities and communities. As part of the many governance indicators (McConaghy 2012), we note the importance of creating a peaceful living environment that entails the core value of the environment. Many aspects would indeed lead to creating prosperity (Benkler and Masum 2008) and enhancing the environmental policy and governance that are the backbone of creating and maintaining healthy environments. Furthermore, as the World Health Organization (2019) highlights, we require to have health in all policies, enabling us to implement the sustainable development agenda through good governance. This means that we need to provide the necessary support to our government and societies, aiming to have a well-functioning environment first before achieving the health and well-being of society.

Anglin (2004) noted that place plays a very important part in a system of care, which also applies to what we care about or neglect in our living environments. The environment that builds the structure and essence of our cities and communities are robust spaces, which still require our attention and continuous care. In the global environment, we see global agendas and motives encouraging more sustainable pathways than unsustainable models that still persist today. The role of our dynamic global economy cannot be neglected in how the global environment is shaped (Aidt and Tzannatos 2002), especially in the age of globalisation where governmental policies are connected with a larger body of international agenda, policies, agreements, etc. However, in all cases, we see the importance of well-functioning regulatory and governments (Agrawal, 2010) in maintaining a healthy environment. This also means that we acquire stability and allow structures of the government and control (Rose-Ackerman 2008) to help to

A. Cheshmehzangi, *Urban Health, Sustainability, and Peace in the Day the World Stopped*, Sustainable Development Goals Series, https://doi.org/10.1007/978-981-16-4888-5_6

improve the living environments and towards peace and sustainability.

As we see around the globe, the trends of having standards and benchmarks have been incredibly helpful in achieving a well-functioning government and dealing with environmental issues (Bagwell and Staiger 2001). Nonetheless, we are perhaps not so successful in many cases and/or are very slow in making any solid progress. Today's focus has been on reducing the impacts or minimising the impacts, and little progress is made on reversing any of the unsustainable trends. With a growing number of natural disasters and increasing global warming issues (Cheshmehzangi and Dawodu 2019), we rest assured that reversing the trends has not been successful to date. The unsustainable pathways are yet to be slowed down in order for us to ensure peaceful and healthy living environments are achieved, scaled up, and sustained.

In most cases, we forget the close relationship that exists between environmental policy, government, and the market (Hepburn 2010). Hence, the by-default thinking would be to put a well-functioning government as part of the institutional dimension of sustainability. However, we beg to differ ad agree with those who suggest a greener revolution may require comprehensive environmental governance (Agrawal and Lemos 2007). Therefore, the interplay of well-functioning government and peaceful living environments could indicate why environmental impact assessments are important, what should be considered in our environmental services, and how we may consider better environmental management from the perspective of governance.

In this chapter, we argue the above points to ensure environmental governance is a precursor to a well-functioning government and vice versa. By achieving a well-functioning government, we can have a well-functioning environment. And if so, then we may be able to consider changes and challenges, especially from the environmental management perspectives (He et al. 2012). The ultimate goals should be—if they are not to date—to have a successful life and a well-functioning society (Gilomen 2003), which cannot be separated from the idea of having peaceful living environments. By referring to living environments, we are suggesting the environments just for us, the humans, and the others in our ecosystem services that our lives depend on, without a doubt. A well-functioning environment, if governed adequately, would then raise our pleasure (Nyakang'o and Bruggen, 1999) and level of peace and satisfaction. This was proven to be very accurate during the COVID-19 pandemic, as our appreciation of the natural environments and quality of life (Cheshmehzangi 2020a) became more apparent. In this regard, we ought to know the value of the well-functioning urban ecosystems (Wit et al. 2012), before we could argue for a well-functioning government. To achieve the former, we need to sort out the latter, thus the importance of environmental management and governance in maintaining peace and sustainability in cities and communities.

6.2 Reflecting on How Cities Reacted During the Pandemic

In some of the earlier work on COVID-19 and environmental sustainability (Arora and Mishra 2020), we highlight some excellent examples that explored possible environmental effects on the spread of the disease (Xu et al. 2020). In their studies, Arora and Mishra (2020) argue the close link between environmental issues and the potential spread of pandemics across the work. They suggest "*global warming and climate change, extreme levels of pollutants in the environment (soil, water, and air), deforestation, fragmentation of natural environments, intensive farming, and globalization are the factors contributing to the emergence and spread of the new wave of deadly pandemics*" (ibid, p. 118). The belief that the closer we are getting to intact natural environments suggests the cause of getting close with diseases in other species. Such a situation triggers the opportunity to pass new diseases from animals to humans. More importantly, two other potential scenarios exist, which suggest the relevance of co-existence with natural habitats. In one example, suggestions are focused on the impact of global warming on potential hidden diseases and how they are releasing into our living environments. The other example also indicates the possibility of us disturbing the natural habitats and creating an imbalance in the natural environments. The former is related to evolution theory as a result of global warming (Galam 2010), and the latter argues that pandemics like the COVID-19 are a sort of Anthropocene disease (O'Callaghan-Gordo and Anto 2020). This relates to the overarching arguments that shrinking natural habitats and changing behaviour are the cause of new disease emergence (Vidal 2020). In other words, a possibility for a consequence of the anthropogenic biodiversity crisis and climate change issues (Lorentzen et al. 2020) may lead to a new pandemic in the future. Based on the evidence that shows 70% of the Earth's natural ecosystem is disturbed (Diaz and Brondizio 2019), the emergence of pandemics cannot be a surprise. What is more, we also see other issues that threaten conservation and the health of natural environments (Epstein et al. 2003), adding to vulnerabilities of our environments and pressuring their management and governance.

Some other studies evaluated various relationships of the pandemic to climate and environmental conditions (Kroumpouzos et al. 2020). New ideas are emerging regarding restructuring ecological and environmental governance (Ryan 2020) and identifying new processes for achieving well-functioning governments. Earlier on, scholars like Bodin (2017) stressed the importance of achieving collective action in social–ecological systems, but we are truly far from that ideal fascination. The approach to potential integrative environmental governance (Visseren-Hamakers 2015) would be something that cities could push to enhance governance and meta-governance. Nonetheless, during the COVID-19 pandemic, cities reacted very differently. In fact, the focus on dealing with COVID-19 calamity led to other vulnerabilities that may reverse some of the recent trends on environmental governance.

In achieving peaceful living environments, we note that a well-functioning government is a necessity. The situation that led to the closure and reduction of productions was first seen as a potential transformation but was then became just a temporary phase. The lack of well-functioning governments across the globe shows that cities and regions are still weak in terms of managing, protecting, and healing their natural environments. The pressure on environments remains and continues as the combination of land and market could potentially make cities to be environmentally vulnerable. For example, during the COVID-19 pandemic, cities may have temporarily experienced better air quality, but this was far from any restoration opportunity. The short-lived conditions led to longer-term issues (Cheshmehzangi 2020b), such as higher dependency on private cars, increasing tourism and entertainment, and a higher level of energy demand and consumption. In addition, we have a higher level of impact on the environment and society (Verma and Prakash 2020), knowing that the impacts have been significant (McNeely 2021).

Suppose we want to face reality accurately. In that case, we realise cities and regions did not react well from the environmental perspective. They did not take the opportunity to reflect on some of their main environmental issues. The absence of a well-functioning government indicates a lack of environmental governance, poorer environmental management, and a lack of monitory mechanisms. The impacts are seen on lack of monitory on carbon caps, emissions, deforestation, environmental protection, land conversion, etc. The idea to call for "*harmonisation of sustainability*" as discussed by Adams and Abhayawansa (2021) suggests the overarching opportunity for a better response to the pandemic through better governance (Sharma et al. 2021). The situation, however, highlights the fact that the pandemic exacerbated existing inequalities (Auld and Renckens 2021), undermining the opportunity for agile and adaptive governance (Janssen and Voort 2020). We see a lack of environmental responsibility, which could indeed lead to longer-term adversities and challenges for cities to manage the impacts. Thus, the many failures should be looked at more smartly, allowing for interventions for better institutional–environmental nexus. The ultimate goal would be to develop and sustain a well-functioning government that addresses environmental issues adequately and promotes environmental governance as its first step in making healthier and peaceful living environments.

6.3 Addressing the Flaws

We could perhaps learn more than what we have experienced during the COVID-19 pandemic. For instance, we have to learn from the pandemic to have better public health climate preparedness (Sheehan and Fox 2020), which would be essential to enhance our resilience against other emergencies and natural disasters. For us to reflect appropriately on the mentioned flaws, we have to consider two directions of adaptive thinking and resilience enhancement (Cheshmehzangi 2020a). The two can only happen through a well-functioning government with core indicators and strategies on environmental governance, environmental resilience, environmental management, and environmental protection. The healthiness and peace in our living environments of all kinds very much depend on the balance of co-existence, which requires a holistic revision and nothing less. The COVID-19 pandemic provides us with learning opportunities for better environmental management, which should help enhance our regional and larger-scale governance beyond the boundaries of cities and communities. The situation highlights weakening institutional–environmental interchanges, which should be more progressive towards achieving the sustainable development agenda and goals.

Unquestionably, cities function more efficiently and reflectively with better management and well-functioning governments. As part of the positive peace pillars, a well-functioning government reflects on multiple factors that involve society (or people), environment, and governance. As mentioned earlier, the apparent focus on the institutional dimension would minimise environmental governance and the environmental agenda in achieving healthy and peaceful living environments. As we deal with the health and economic effects of the pandemic on societies, we should not neglect environmental sustainability in maintaining and sustaining peace in our cities. Such ignorance may simply lead to more prominent effects on waste management, environmental protection, and our investment in environmental sustainability.

Suppose we opt to disconnect a well-functioning government from environmental governance. In that case, we may neglect the importance of built-natural nexus, ecosystem services of cities and their surrounding areas, and higher-level strategies aimed to minimise our impacts on the environments and govern them more sustainably. Our goal should be to create peaceful living environments, have healthier synergies between cities and natural habitats, reduce the disturbances that cause potential disorders and environmental disasters, and help maximise our ecologies' benefits, including the urban ecologies surrounding us. Cities and communities should call for well-functioning governments by putting environmental agenda and governance as their priority. Without the two, we are stuck with the never-ending cycle of economic growth, straying away from healthy living and environmental sustainability.

6.4 A Summary

In this chapter, we argued in favour of the institutional-environmental nexus to achieve a well-functioning government. If we consider this nexus punctually, peaceful living environments could be guaranteed. But if not, we may destabilise the balance that shapes our co-existence with nature. In our earlier work, we discussed whether humans are cure or curse in responding to environmental and climate crises (Cheshmehzangi and Dawodu 2019). We demonstrated examples of how they can still become the cure. The situation persists under the shadow of a prolonged pandemic that has caused larger-scale adversities, inequalities, and inactions around the globe. A well-functioning government stands as a significant pillar that connects many aspects of environmental sustainability, some that require institutional support and some that may need revision and reconsideration.

In the following two chapters, we highlight two other overarching aspects under the environmental sustainability dimension. For us, the relationship between these aspects and different dimensions is apparent. Therefore, it is important that we highlight—once again—the position of health–sustainability–peace (HSP) nexus in achieving healthy environments for healthier cities and people.

References

Adams CA, Abhayawansa S (2021) Connecting the COVID-19 pandemic, environmental, social and governance (ESG) investing and calls for 'harmonisation' of sustainability reporting. Crit Perspect Acc 102309. https://doi.org/10.1016/j.cpa.2021.102309

Agrawal A (2010) Environment, community, government. In the name of humanity. Duke University Press, Durham, NA, pp 190–217

Agrawal A, Lemos MC (2007) A greener revolution in the making?: environmental governance in the 21st century. Environ: Sci Policy Sustain Dev 49(5):36–45

Aidt T, Tzannatos Z (2002) Unions and collective bargaining: economic effects in a global environment. World Bank, Washington, DC

Anglin JP (June 2004) Creating "well-functioning" residential care and defining its place in a system of care. In: Child and youth care forum (vol 33, no. 3). Kluwer Academic Publishers-Plenum Publishers, pp 175–192

Arora NK, Mishra J (2020) COVID-19 and importance of environmental sustainability. Environ Sustain 3:117–119. https://doi.org/10.1007/s42398-020-00107-z.pdf

Auld G, Renckens S (2021) Private sustainability governance, the Global South and COVID-19: are changes to audit policies in light of the pandemic exacerbating existing inequalities? World Dev 139:105314

Bagwell K, Staiger RW (2001) The WTO as a mechanism for securing market access property rights: implications for global labor and environmental issues. J Econ Perspect 15(3):69–88

Benkler Y, Masum H (2008) Collective intelligence: Creating a prosperous world at peace. Earth Intelligence Network, Oakton

Bodin Ö (2017) Collaborative environmental governance: achieving collective action in social-ecological systems. Science 357(6352)

Cheshmehzangi A (2020a) The city in need: urban resilience and city management in disruptive disease outbreak events. Springer, Singapore

Cheshmehzangi A (2020b) COVID-19 and household energy implications: what are the main impacts on energy use? Heliyon 6(10):e05202

Cheshmehzangi A, Dawodu A (2019) Sustainable urban development in the age of climate change—people: the cure or curse. Palgrave Macmillan, Singapore

de Wit M, van Zyl H, Crookes D, Blignaut J, Jayiya T, Goiset V, Mahumani B (2012) Including the economic value of well-functioning urban ecosystems in financial decisions: evidence from a process in Cape Town. Ecosyst Serv 2:38–44

Diaz SSJ, Borondizio E (2019) Summary for policymakers of the global assessment report on biodiversity and ecosystem services of the Intergovernmental Science-Policy Platform on Biodiversity and Ecosystem Services. IPBES/7/10/Add1, pp 1–45

Epstein PR, Chivian E, Frith K (2003) Emerging diseases threaten conservation. Environ Health Perspect 111 (10):A506–A507

Galam S (2010) Public debates driven by incomplete scientific data: the cases of evolution theory, global warming and H1N1 pandemic influenza. Phys A 389 (17):3619–3631

Gilomen H (2003) Desired outcomes: a successful life and a well-functioning society. Key competencies for a successful life and a well-functioning society, pp 109–134

He G, Lu Y, Mol AP, Beckers T (2012) Changes and challenges: China's environmental management in transition. Environ Dev 3:25–38

Hepburn C (2010) Environmental Policy, Government, and the Market, Oxford Review of Economic Policy, 26(2):117–136

Janssen M, van der Voort H (2020) Agile and adaptive governance in crisis response: lessons from the COVID-19 pandemic. Int J Inf Manage 55:102180

Kroumpouzos G, Gupta M, Jafferany M, Lotti T, Sadoughifar R, Sitkowska Z, Goldust M (2020) COVID-19: a relationship to climate and environmental conditions?. Dermatologic therapy, p E13399. https://www.ncbi.nlm.nih.gov/pmc/articles/PMC7235510/

Lorentzen HF, Benfield T, Stisen S, Rahbek C (2020) COVID-19 is possibly a consequence of the anthropogenic biodiversity crisis and climate changes. Dan Med J 67(5):A205025

McConaghy C (2012) The global peace index and the structure of peace. In: Cooperation for a peaceful and sustainable world part 1. Emerald Group Publishing Limited

McNeely JA (2021) Nature and COVID-19: the pandemic, the environment, and the way ahead. Ambio 1–15

Nyakang'o JB, Van Bruggen JJA (1999) Combination of a well functioning constructed wetland with a pleasing landscape design in Nairobi, Kenya. Water Sci Technol 40(3):249–256

O'Callaghan-Gordo C, Antó JM (2020) COVID-19: the disease of the anthropocene. Environ Res. https://www.ncbi.nlm.nih.gov/pmc/articles/PMC7227607/

Rose-Ackerman S (2008) Corruption and government. Int Peacekeeping 15(3):328–343

Ryan E (2020) Lessons from the Coronavirus pandemic for environmental governance. Seeing the woods: a Rachel Carson center blog. https://papers.ssrn.com/sol3/papers.cfm?abstract_id=3648648

Rychen DS, Salganik LH (eds) (2003) Key competencies for a successful life and well-functioning society. Hogrefe Publishing

Sharma A, Borah SB, Moses AC (2021) Responses to COVID-19: the role of governance, healthcare infrastructure, and learning from past pandemics. J Bus Res 122:597–607

Sheehan MC, Fox MA (2020) Early warnings: the lessons of COVID-19 for public health climate preparedness. Int J Health Serv 50(3):264–270

Verma A, Prakash S (2020) Impact of covid-19 on environment and society. J Global Biosci 9(5):7352–7363

Vidal J (2020) Destroyed habitat creates the perfect conditions for coronavirus to emerge. Scientific American, p 18

Visseren-Hamakers IJ (2015) Integrative environmental governance: enhancing governance in the era of synergies. Curr Opin Environ Sustain 14:136–143

World Health Organization (2019) Adelaide statement II on health in all policies: implementing the sustainable development agenda through good governance for health and well-being: building on the experience of health in all policies (No. WHO/CED/PHE/SDH/19.1). World Health Organization

Xu H, Yan C, Fu Q, Xiao K, Yu Y, Han D, Wang W, Cheng J (2020) Possible environmental effects on the spread of COVID-19 in China. Sci Total Environ 731:139211

7 The City Needs Its Essential Systems: Sustaining the Primary Operations

Chasing the light is not a necessity, but losing it will be extremely harmful.
—Ali Cheshmehzangi, 2018

7.1 A General Overview of Urban Essential Systems from the Environmental Sustainability Perspective

As a complementary section to Chap. 5, we look into some essential urban systems from the environmental sustainability perspective. They usually belong to specific categories under the critical infrastructures (CIs). Some of these include waste management, water management, sewage system management, wastewater management, etc. A recent study looks into plastic waste, energy, and environmental footprints related to COVID-19 (Klemeš et al. 2020). In their research, both economic and environmental issues are highlighted, and further discussions are made on waste management systems, especially for contaminated waste during COVID-19. Our earlier work also verifies larger waste production that adds to the existing capacity of waste management systems (Cheshmehzangi 2020a), showing the neglect in dealing with such critical environmental issues. There are also examples of evaluating indirect effects of COVID-19 on the environment (Zambrano-Monserrate et al. 2020), such as those related to water, sanitation, hygiene, and waste management issues. More specifically, food systems sustainability is questioned (ibid), which again responds to some of the key essential public services (Cheshmehzangi 2020b) and how cities may respond to situations like health emergencies.

In many cases, the utilisation of waste as a resource has not been well implemented or managed (Dawodu et al. 2021). This means that we still lack a good understanding of how waste is managed, where it goes, and how it harms the environment. If you ask the general public, they may be aware of the situation but may not know how the process occurs or where the waste ends up in most cases. Moreover, during the COVID-19, we see further issues on inequality (Ashford et al. 2020), leading to lower quality management of essential urban systems and larger impacts on the immediate environments. Some suggest the need for restructuring society and the environment (Elavarasan and Pugazhendhi 2020), suggesting ways of reducing the environmental effects and or characterising how essential systems could be better managed in terms of their operations and usage. In the light of such ideas, we see growing concerns in regard to unsustainable waste management systems (You et al. 2020). This shows there are sector-based environmental concerns (Kumar et al. 2020) and potential influence on certain aspects

A. Cheshmehzangi, *Urban Health, Sustainability, and Peace in the Day the World Stopped*, Sustainable Development Goals Series, https://doi.org/10.1007/978-981-16-4888-5_7

such as waste generation (Fan et al. 2021), biomedical waste (Rahman et al. 2020), and a larger-scale threat to the marine environment (Dharmaraj et al. 2021). In return, there are examples of challenges and strategies for specific areas of addressing the effectiveness of our essential systems, particularly in cities. For instance, Vanapalli et al. (2021) highlight plastic waste management strategies, which again related to major factors of inclusive waste governance and solid waste management plans (Nzeadibe and Ejike-Alieji 2020). These approaches or strategies are essential to overcome the long-lasting issues of solid waste, concerning matters that could lead to lower quality hygiene and polluted environments through solid waste and other waste.

The existing literature provides us with a large body of evidence that cities had significant issues related to water and wastewater management issues, which led to significant public health and geo-environment concerns (Annalaura et al. 2020; Daniell 2020; Paleologos et al. 2020). In addition, we see that issues of water resource management were highlighted well (Neal 2020), showing that "*COVID-19 has exposed and deepened the inequalities that exist*". These studies evaluate the nexus between water and the environment, considering the impacts of such crises or emergencies on our essential systems, leading to water sanitation and hygiene challenges. As highlighted by the World Health Organization (2020), we need to consider technical processes in dealing with such issues, especially during the pandemic events like the COVID-19. On one side, we are dealing with widening natural environmental impacts (Rupani et al. 2020). On the other side, we see weakening environmental governance for managing waste, water, wastewater, sewage systems, etc. The latter puts further pressure on the former, knowing that our digital revolution or progression (Poch et al. 2020) may not be necessarily enough to deal with the environmental impact of the pandemics. Many studies (Aydin et al. 2020; Barbier and Burgess 2020; Rizou et al. 2020; Somani et al. 2020; You et al. 2020; Tiwari et al. 2021) suggest similar effects or implications of COVID-19 on the environment. Some of these studies raise their concerns on water safety issues from water management perspectives, water treatment methods, risk management practices, WASH (Water, Sanitation, and Hygiene), and waste management in and around cities.

Suppose we opt to deal with such issues effectively. In that case, we have to consider the use of sustainable urban waste management systems (Hellström et al. 2000) and better preparedness of our essential systems dealing directly and indirectly with the environmental sustainability of cities and regions. This chapter explores some of these essential systems (similar to those in Chap. 5), as they are vital to sustaining the city's healthy operations. In achieving these, cities could maintain healthier environments for people and other habitats. We also need to seek better methods of waste reduction, waste management, water and wastewater management, water treatment, and aim to increase the levels of sanitation and hygiene across cities and communities.

7.2 Reflecting on How Cities Reacted During the Pandemic

A significant aspect of preventing the disease transmission was surgical gloves, added medical supplies, and sanitisers (Saadat et al. 2020). However, the situation led to larger-scale waste production, leading to the sudden increase of medical waste in the environment. Richter et al. (2021) also identified changes in waste disposal characteristics in Canadian cities, showing specific changes and effects on waste management issues. While cities such as Wuhan, the first epicentre of the COVID-19 in China, were appraised for their success in waste management strategies (Singh et al. 2020), many other cities faced difficulties in dealing with the added waste and keeping the essential systems safe and operational.

One of the suitable examples of public health tools is the National Wastewater Surveillance System (NWSS), which has been initiated in response to the COVID-19 pandemic. The new tool is a collaboration between the US Department of Health and Human Services (HHS), the Centers for Disease Control and Prevention (CDC), and associated agencies throughout the federal government (CDC, 2021). In dealing with wastewater (also known as sewage) systems, the tool uses sample data that studies household waste, water, and non-household sources, including rainwater and industrial use. In doing so, the tool enables better sewage testing through sampling and helping to do early detection of diseases, which is commonly practised for other pandemics. Moreover, wastewater surveillance could enable tracking and assessing sources and help to reduce infection distribution through essential urban systems. The information on such systems in urban areas is provided openly, which addresses the availability of free flow information. It also encourages engagement of other stakeholders or local partners to ensure dealing with wastewater treatment processes, assessment, reporting, and identifying what should be done in action plans. Given the opportunity to utilise such tools widely, we can enhance our public health action plan and respond more effectively to community-level wastewater issues, treatment, and management. Similar practices or strategies are also developed in Wisconsin, the USA, where a wastewater monitoring network is developed to provide high-quality data to all (Wisconsin Department of Health Services 2021). The method is effective for early detection and better management of the spread within communities:

> Monitoring wastewater is used for early detection of COVID-19 within a community. For people with COVID-19, the virus can be detected in their faeces shortly after they are infected with the virus, even before they experience symptoms or if they are infected but asymptomatic. By testing wastewater, we can measure the amount of the virus and see whether the levels are increasing or decreasing. This can be an early warning sign of increasing COVID-19 cases within a community (ibid).

Another example is the wastewater COVID-19 tracking strategy, which developed a pilot study by the Massachusetts Water Resource Authority (MWRA 2021), which collaborates with a wastewater epidemiology company. In this example, collection, tracking, and monitory approaches are used again. This approach is now commonly practised for tracking the disease with wastewater (Larsen and Wigginton 2020), indicating that the concentration of the virus in primary sewage systems (Peccia et al. 2020) could be a problematic situation for cities and communities. Similar tools or strategies are applied elsewhere (Daughton 2020; Farkas et al. 2020; Polo et al. 2020; Randazzo et al. 2020; Street et al. 2020; Wang et al. 2020), highlighting specific areas such as wastewater risks and monitory (Bogler et al. 2020), early detection analysis (Venugopal et al. 2020), and treating contamination of wastewater by the disease (Lahrich et al. 2021). In these examples, we see major concerns that could lead us to enhance our waste and wastewater services sector (Nghiem et al. 2020), which will be beneficial for the post-pandemic era.

Furthermore, some strategies respond to a specific context or cultural demand. In all these cases, water treatment is an important factor in ensuring that the health and peace of communities are maintained. For instance, having a separate water treatment after cleansing the dead bodies with the COVID-19 infection in Iran helped to change common procedures but keep the cultural and/religious rituals (Acrofan 2021; Nasseri et al. 2021). Other studies or interventions suggest scope for innovations for effective solid waste management (Sharma et al. 2020; Filimonau 2021). Improper solid waste management is highlighted as an increasing problem (Nzediegwu and Chang 2020), particularly in locations where urban systems are weak and/or where sanitation levels are relatively low. As shown in some examples, the same issues persist in food waste management (Aldaco et al. 2020), increasing household consumption (Cheshmehzangi 2020c), and biomedical waste (Ilyas et al. 2020). While we have two sides of the spectrum of thriving cities with robust strategies and adaptive models, we also note many cities have deficiencies in addressing the mentioned environmental effects. The following section highlights some of the critical suggestions to ensure environmental sustainability is maintained through sustaining the primary operations of cities.

7.3 Addressing the Flaws

As highlighted in the examples of the previous section, some practical innovations could help track and monitor the spread of the virus in our waste and water systems. Nonetheless, there is little information about how we could safeguard those systems in the first place. In many cases, untreated wastewater could lead to complexities, which could also affect communities (Ahmed et al. 2020). Therefore, there is a potential health risk (Lodder and Husman 2020) that cannot be simply neglected after the COVID-19 pandemic is over. In addition, we should be able to have a better waste management system, enabling us to separate waste more systematically and effectively. For instance, separating medical waste from general waste could be a strategy that should be implemented globally. Nevertheless, in more impoverished communities, such processes are not in place, and waste collection still happens informally. Therefore, there are higher risk levels in contexts where solid waste is not managed through formal procedures and where contaminated waste is treated indifferently.

For wastewater treatment and management, as Kataki et al. (2020) suggest, there are two sets of concerns, including concern during treatment and after treatment. In these processes, we have to consider treatments of chemical, physical, and chemical–physical waste (ibid), highlighting residual waste as well as minimising the non-sanitised environments and potential risks through unsustainable management. In this regard, the waste and wastewater services sectors require revisiting, which should enable us to integrate better treatment systems, more holistic management systems, and treatment technologies. As Tiwari et al. (2021) highlight, environmental and health security perspectives are essential in addressing the sustainability of our living environments. Therefore, we must consider the implications for water and waste sectors more cautiously (Hill et al. 2020) and develop national response plans, regional/local initiatives, and community-level interventions.

The flaws in such essential urban systems may not be the same in each context. Again, we see growing inequalities as some cities with older infrastructures have struggled in managing their wastewater and waste. The application of recycled wastewater irrigation (Oliver et al., 2020) and immediate response plan to waste management (Singh et al., 2020) may not be necessarily applicable to all contexts. Therefore, cities have to evaluate context-specific factors more cautiously, and decision-makers should respond to them more responsibly.

7.4 A Summary

As shown in this chapter, our essential urban systems, mainly from the environmental sustainability consideration, are critical to the healthy operations of cities. By sustaining such primary operations, we can also enhance the city's resilience and public health (Cheshmehzangi 2020d), which will also be critical to cities that could respond to pandemics/outbreaks. As shown in examples of restructuring future cities (Chen et al. 2020), we see new opportunities emerging that suggest integrating new technologies in current practice and maintaining community well-being. These approaches should lead us towards future urban health policies (Allam et al. 2020) focused mainly on instrumental mechanisms in safeguarding society and not just urban economies.

Based on the current scenarios, we see significant differences between cities with well-developed infrastructures and those lacking basic infrastructures. The examples of waste management, water management, wastewater management, and similar factors are just a few examples of essential urban systems that require more attention from the environmental sustainability perspective. In order to maintain a good balance in the health–sustainability–peace (HSP) nexus, we ought to consider the value of institutional–environmental and technical–environmental combinations. The more recent changes in sustainability priorities in organisations (Barreiro-Gen et al. 2020) would potentially be problematic, especially if socio-economic issues override the long-lasting environmental crises in cities and the built environments. Such changes may not be so helpful for revisiting the management of essential urban systems, particularly those that were covered throughout this chapter. As evidenced during the COVID-19 pandemic, much of our issues originate from lack of care about our environments, specifically those essential to our operations in cities and communities. As we continuously put pressure on such aspects, we have to reflect more responsibly on how we could resolve the long-term issues of environmental sustainability in cities. These factors also lead to further pressures on natural habitats and resources, mainly from the environmental perspective. To follow up more specifically on these points, the next chapter delves into other discussions on resources. The narrative will be mainly from the environmental sustainability perspective, once again addressing what is still missing from the critical environmental dimension.

References

Acrofan (2021) The 5th Guangzhou award shortlist released in Guangzhou. https://us.acrofan.com/detail.php?number=416907

Ahmed W, Angel N, Edson J, Bibby K, Bivins A, O'Brien JW, Choi PM, Kitajima M, Simpson SL, Li J, Tscharke B (2020) First confirmed detection of SARS-CoV-2 in untreated wastewater in Australia: a proof of concept for the wastewater surveillance of COVID-19 in the community. Sci Total Environ 728:138764

Aldaco R, Hoehn D, Laso J, Margallo M, Ruiz-Salmón J, Cristobal J, Kahhat R, Villanueva-Rey P, Bala A, Batlle-Bayer L, Fullana-I-Palmer P (2020) Food waste management during the COVID-19 outbreak: a holistic climate, economic and nutritional approach. Sci Total Environ 742:140524

Allam Z, Dey G, Jones DS (2020) Artificial intelligence (AI) provided early detection of the coronavirus (COVID-19) in China and will influence future urban health policy internationally. AI 1(2):156–165

Annalaura C, Ileana F, Dasheng L, Marco V (2020) Making waves: coronavirus detection, presence and persistence in the water environment: state of the art and knowledge needs for public health. Water Res 115907

Ashford NA, Hall RP, Arango-Quiroga J, Metaxas KA, Showalter AL (2020) Addressing inequality: the first step beyond COVID-19 and towards sustainability. Sustainability 12(13):5404

Aydın S, Nakiyingi BA, Esmen C, Güneysu S, Ejjada M (2020) Environmental impact of coronavirus (COVID-19) from Turkish perceptive. Environ, Dev Sustain 1–8

Barbier EB, Burgess JC (2020) Sustainability and development after COVID-19. World Dev 135:105082

Barreiro-Gen M, Lozano R, Zafar A (2020) Changes in sustainability priorities in organisations due to the COVID-19 outbreak: averting environmental rebound effects on society. Sustainability 12(12):5031

Bogler A, Packman A, Furman A, Gross A, Kushmaro A, Ronen A, Dagot C, Hill C, Vaizel-Ohayon D, Morgenroth, Bertuzzo E (2020) Rethinking wastewater risks and monitoring in light of the COVID-19 pandemic. Nature Sustain 1–10

Centers for Disease Control and Prevention (CDC) (2021) National wastewater surveillance system (NWSS). https://www.cdc.gov/coronavirus/2019-ncov/cases-updates/wastewater-surveillance.html

Chen B, Marvin S, While A (2020) Containing COVID-19 in China: AI and the robotic restructuring of future cities. Dialogues Human Geogr 10(2):238–241

Cheshmehzangi A (2020a) Introduction: The city during outbreak events. In: Cheshmehzangi A (ed) The city in need: urban resilience and city management in disruptive disease outbreak events. Springer, Singapore, pp 1–15

Cheshmehzangi A (2020b) Responsiveness through city management. In: Cheshmehzangi A (ed) The city in need: urban resilience and city management in disruptive disease outbreak events. Springer, Singapore, pp 105–135

Cheshmehzangi A (2020c) COVID-19 and household energy implications: what are the main impacts on energy use? Heliyon 6(10):e05202

Cheshmehzangi A (2020d) Reflection on early lessons for urban resilience and public health enhancement during the COVID-19. Health 12(10):1390

Daniell KA (2020) Water management beyond the fortressed COVID-19 world: considerations for the long-term. Australas J Water Resour 24(2):85–90

Daughton CG (2020) Wastewater surveillance for population-wide Covid-19: the present and future. Sci Total Environ 139631

Dawodu A, Oladejo J, Tsiga Z, Kanengoni T, Cheshmehzangi A (2021) Underutilization of waste as a resource: bottom-up approach to waste management and its energy implications in Lagos, Nigeria. Intell Build Int 1–22

Dharmaraj S, Ashokkumar V, Hariharan S, Manibharathi A, Show PL, Chong CT, Ngamcharussrivichai C (2021) The COVID-19 pandemic face mask waste: a blooming threat to the marine environment. Chemosphere 272:129601

Elavarasan RM, Pugazhendhi R (2020) Restructured society and environment: a review on potential technological strategies to control the COVID-19 pandemic. Sci Total Environ 725:138858

Farkas K, Hillary LS, Malham SK, McDonald JE, Jones DL (2020) Wastewater and public health: the potential of wastewater surveillance for monitoring COVID-19. Curr Opin Environ Sci Health

Filimonau V (2021) The prospects of waste management in the hospitality sector post COVID-19. Resour, Conserv, Recycl 168:105272

Hellström D, Jeppsson U, Kärrman E (2000) A framework for systems analysis of sustainable urban water management. Environ Impact Assess Rev 20(3):311–321

Hill K, Zamyadi A, Deere D, Vanrolleghem PA, Crosbie N (2020) SARS-CoV-2 known and unknowns, implications for the water sector and wastewater-based epidemiology tosupport national responses worldwide: early review of global experiences with the COVID-19 pandemic. Water Qual Res J. https://doi.org/10.2166/wqrj.2020.100

Ilyas S, Srivastava RR, Kim H (2020) Disinfection technology and strategies for COVID-19 hospital and bio-medical waste management. Sci Total Environ 749:141652

Kataki S, Chatterjee S, Vairale MG, Sharma S, Dwivedi SK (2020) Concerns and strategies for wastewater treatment during COVID-19 pandemic to stop plausible transmission. Resour Conserv Recycl 105156

Klemeš JJ, Van Fan Y, Tan RR, Jiang P (2020) Minimising the present and future plastic waste, energy and environmental footprints related to COVID-19. Renew Sustain Energy Rev 127:109883

Kumar V, Singh SB, Singh S (2020) COVID-19: Environment concern and impact of Indian medicinal system. J Environ Chem Eng 104144

Lahrich S, Laghrib F, Farahi A, Bakasse M, Saqrane S, El Mhammedi MA (2021) Review on the contamination of wastewater by COVID-19 virus: impact and treatment. Sci Total Environ 751:142325

Larsen DA, Wigginton KR (2020) Tracking COVID-19 with wastewater. Nat Biotechnol 38(10):1151–1153

Lodder W, de Roda Husman AM (2020) SARS-CoV-2 in wastewater: potential health risk, but also data source. Lancet Gastroenterol Hepatol 5(6):533–534

Massachusetts Water Resources Authority (MWRA) (2021) Wastewater COVID-19 tracking. https://www.mwra.com/biobot/biobotdata.htm

Nasseri S, Yavarian J, Baghani AN, Azad TM, Nejati A, Nabizadeh R, Alimohammadi M et al (2021) The presence of SARS-CoV-2 in raw and treated wastewater in 3 cities of Iran: Tehran, Qom and Anzali during coronavirus disease 2019 (COVID-19) outbreak. J Environ Health Sci Eng 1–12

Nghiem LD, Morgan B, Donner E, Short MD (2020) The COVID-19 pandemic: considerations for the waste and wastewater services sector. Case Stud Chem Environ Eng 1:100006

Neal MJ (2020) COVID-19 and water resources management: reframing our priorities as a water sector. Water Int 45(5):435–440

Nzeadibe TC, Ejike-Alieji AU (2020) Solid waste management during Covid-19 pandemic: policy gaps and prospects for inclusive waste governance in Nigeria. Local Environ 25(7):527–535

Nzediegwu C, Chang SX (2020) Improper solid waste management increases potential for COVID-19 spread in developing countries. Resour, Conserv, Recycl 161:104947

Oliver MMH, Hewa GA, Pezzaniti D, Haque MA, Haque S, Haque MM, Moniruzzaman M, Rahman MM, Saha KK, Kadir MN (2020) COVID-19 and recycled wastewater irrigation: a review of implications

Paleologos EK, O'Kelly BC, Tang CS, Cornell K, Rodríguez-Chueca J, Abuel-Naga H, Koda E, Farid A, Vaverková MD, Kostarelos K, Goli VSNS (2020) Post Covid-19 water and waste water management to protect public health and geoenvironment. Environ Geotech 40:1–15. https://doi.org/10.1680/jenge.20.00067

Peccia J, Zulli A, Brackney DE, Grubaugh ND, Kaplan EH, Casanovas-Massana A, Ko AI, Malik AA, Wang D, Wang M, Warren JL (2020) Measurement of SARS-CoV-2 RNA in wastewater tracks community infection dynamics. Nat Biotechnol 38(10):1164–1167

Poch M, Garrido-Baserba M, Corominas L, Perelló-Moragues A, Monclús H, Cermerón-Romero M, Melitas N, Jiang SC, Rosso D (2020) When the fourth water and digital revolution encountered COVID-19. Sci Total Environ 744:140980

Polo D, Quintela-Baluja M, Corbishley A, Jones DL, Singer AC, Graham DW, Romalde JL (2020) Making waves: Wastewater-based epidemiology for COVID-19–approaches and challenges for surveillance and prediction. Water Res 186:116404

Rahman MM, Bodrud-Doza M, Griffiths MD, Mamun MA (2020) Biomedical waste amid COVID-19: perspectives from Bangladesh. Lancet Glob Health

Randazzo W, Cuevas-Ferrando E, Sanjuán R, Domingo-Calap P, Sánchez G (2020) Metropolitan wastewater analysis for COVID-19 epidemiological surveillance. Int J Hyg Environ Health 230:113621

Richter A, Ng KTW, Vu HL, Kabir G (2021) Waste disposal characteristics and data variability in a midsizedCanadian city during COVID-19. Waste Management 122:49–54

Rizou M, Galanakis IM, Aldawoud TM, Galanakis CM (2020) Safety of foods, food supply chain and environment within the COVID-19 pandemic. Trends Food Sci Technol 102:293–299

Rupani PF, Nilashi M, Abumalloh RA, Asadi S, Samad S, Wang S (2020) Coronavirus pandemic (COVID-19) and its natural environmental impacts. Int J Environ Sci Technol 1–12

Saadat S, Rawtani D, Hussain CM (2020) Environmental perspective of COVID-19. Sci Total Environ 728:138870

Sharma HB, Vanapalli KR, Cheela VS, Ranjan VP, Jaglan AK, Dubey B, Goel S, Bhattacharya J (2020) Challenges, opportunities, and innovations for effective solid waste management during and post COVID-19 pandemic. Resour, Conserv Recycl 162:105052

Singh N, Tang Y, Zhang Z, Zheng C (2020) COVID-19 waste management: effective and successful measures in Wuhan, China. Resour, Conserv, Recycl 163:105071

Somani M, Srivastava AN, Gummadivalli SK, Sharma A (2020) Indirect implications of COVID-19 towards sustainable environment: an investigation in Indian context. Bioresour Technol Rep 11:100491

Street R, Malema S, Mahlangeni N, Mathee A (2020) Wastewater surveillance for Covid-19: an African perspective. Sci Total Environ 743:140719

Tiwari SB, Gahlot P, Tyagi VK, Zhang L, Zhou Y, Kazmi AA, Kumar M (2021) Surveillance of Wastewater for Early Epidemic Prediction (SWEEP): environmental and health security perspectives in the post COVID-19 Anthropocene. Environ Res 195:110831

Van Fan Y, Jiang P, Hemzal M, Klemeš JJ (2021) An update of COVID-19 influence on waste management. Sci Total Environ 754:142014

Vanapalli KR, Sharma HB, Ranjan VP, Samal B, Bhattacharya J, Dubey BK, Goel S (2021) Challenges and strategies for effective plastic waste management during and post COVID-19 pandemic. Science Total Environ 750:141514

Venugopal A, Ganesan H, Raja SSS, Govindasamy V, Arunachalam M, Narayanasamy A, Sivaprakash P, Rahman PK, Gopalakrishnan AV, Siama Z, Vellingiri B (2020) Novel wastewater surveillance strategy for early detection of COVID–19 hotspots. Curr Opin Environ Sci Health

Wang J, Shen J, Ye D, Yan X, Zhang Y, Yang W, Li X, Wang J, Zhang L, Pan L (2020) Disinfection technology of hospital wastes and wastewater: suggestions for disinfection strategy during coronavirus disease 2019 (COVID-19) pandemic in China. Environ Pollut 114665

Wisconsin Department of Health Services (WDHS). (2021). COVID-19: Wisconsin coronavirus wastewater monitoring network. https://www.dhs.wisconsin.gov/covid-19/wastewater.htm

World Health Organization. (2020). Water, sanitation, hygiene and waste management for COVID-19: technical brief, 03 March 2020 (No. WHO/2019-NcOV/IPC_WASH/2020.1). World Health Organization

You S, Sonne C, Ok YS (2020) COVID-19's unsustainable waste management. Science 368(6498):1438

Zambrano-Monserrate MA, Ruano MA, Sanchez-Alcalde L (2020) Indirect effects of COVID-19 on the environment. Sci Total Environ 728:138813

8 Resource Efficiency and Distribution: Healthy Environments for Healthy Cities

We talk, and we observe. That is what makes our world today.

—Ali Cheshmehzangi, 2018

8.1 A General Overview of Healthy Resource Efficiency and Distribution

The continuing climate change crises lead to resource scarcity issues, jeopardising global peacebuilding processes. It also creates larger gaps between different contexts/regions/sub-regions. As Davies-Vengoechea (2004) highlights, "*keeping the peace distribution of resource and goods*" is part of sustaining health and peace in societies. This is also verified by other scholars, like Rissler and Shields (2019), who recognise positive peace as essentiality. Nonetheless, from the positive peace perspective, resource efficiency is generally linked with equity factors. Therefore, the primary factor is regarded as equitable distribution of resources. Here, we aim to look at this from the environmental perspective. More importantly, resource efficiency would link into sustainable consumption and production, generally related to a circular economy (Di Maio et al. 2017; DeWit 2020; Mori et al. 2020). Furthermore, Freire-González and Vivanco (2020) highlight the environmental rebound effect related to the COVID-19 pandemic, specifically looking at environmental and resource management aspects. In this regard, we see greater importance of resource efficiency from the environmental sustainability perspective, responding more direction to discourse and practice of transitions in policymaking studies (Happaerts 2016). From peace-making thinking, we also note that resource efficiency should come with resource distribution factors, which could cultivate peace and provide better approaches for natural resource management (Buckles 1999). In their "*handbook on sustainability transition and sustainable peace*", Brauch et al. (2016) put together some outstanding examples of practices that suggest resource efficiency and sustainable peace nexus as a sort of ideal way in keeping our environments healthy.

Based on the discussions of Ramaswami et al. (2016), a healthy city must have diverse strategies for resource efficiency, some that may then lead to resilience enhancement in urban systems and networks (Cheshmehzangi 2020a, b). Despite the many economic implications of resource efficiency (Ekins et al. 2016), it is a large part of the environmental sustainability of

A. Cheshmehzangi, *Urban Health, Sustainability, and Peace in the Day the World Stopped*, Sustainable Development Goals Series, https://doi.org/10.1007/978-981-16-4888-5_8

healthy cities (Corburn 2009). Some of these are expressed comprehensively in the urban material metabolism and its eco-efficiency (Zhang and Yang 2007), which highlight the importance of self-reliant cities and health promotion in cities and communities (Morris 1987; Duhl et al. 1999; Gutiérrez-Montes 2005; Zhou 2006; Agbali et al. 2017). In this regard, we note the importance of resources from various viewpoints, including—but not limited to—efficiency and distribution. Indeed, the two are also integrated with overarching factors, such as equity and management, to ensure people have access to resources of various sorts. Partly related to the institutional dimension and partially sits with the economic dimension, resources are essential in many ways.

As highlighted by Brock (1991), people often appreciate their (natural) environments as scarce resources, which has been widely accepted in the peace research agenda. More importantly, we refer to the role of natural resource management in achieving sustainable peace (Krampe 2017), which also addresses three more prominent areas of human development, stability of peace and sustainable development. Moreover, resource management of any sort is vital to peace-making and managing peace. Thus, natural resources directly link to communities' affluence and what needs to be protected, accessed, and shared. In many cases, resource scarcity is regarded as a conflict (Bretthauer 2015) against peace processes and sustainable peace, meaning that cities could be divine places with good quality natural resources, environments, and people's free accessibility to them.

In many tropical cities and communities of the global south, we see a good range of natural resources, often lacking proper management. In those cases, we see people's inaccessibility to such resources or certain limitations that do not allow the free flow of people in places that are part of their living environments. More importantly, those natural resources are not adequately shared, leading to inequitable situations that often include conflicts between communities or between communities and their local governments. If we relate these matters to the United Nations' Sustainable Development Goals (SDGs), we can also verify the complexities that exist in non-inclusive communities. For instance, factors such as having access to clean water, clean air, and clean environments are perhaps basic ideas that are still largely missing in many developing countries. The conditions are set in a way that natural resources and environments are still separated from average life. In some cases, significant threats exist in particular to natural resources through the rise of environmental crime (Nellemann et al. 2016), civil conflicts or unrests (Schafer and Black 2003; Marton-Lafevre 2007; Krampe 2017), and lack of education (Harris 2004; Bajaj and Chiu 2009). While the aim to cover resource efficiency and distribution matters from the environmental sustainability perspective, this chapter cannot cover all the above matters. Hence, we aim to mainly focus on natural resources and their management for the health–sustainability–peace (HSP) nexus. Thus, this chapter highlights only a few aspects that would relate to resources in cities from the environmental/natural perspectives.

8.2 Reflecting on How Cities Reacted During the Pandemic

As we see during the rollercoaster events of the COVID-19 pandemic, cities and communities (and many nations) did not react well in regard to resource efficiency and distribution. One thing that was widely recognised was the growing public awareness of nature and the environment during the pandemic (Rousseau and Deschacht 2020). Despite the many environmental impacts, food and natural resources have become a major part of intergenerational equity and towards potential "*behavioural changes to the environment*" (Helm 2020). Such wider recognition led to a better understanding of the human–environment system and connected the two in understanding human society and the natural environment (Cheshmehzangi 2020c; Sarkar et al. 2021). In this regard, topics such as natural resource governance (Walters et al. 2021) became more popular for future directions of human nature harmonious development and better "*resource use and access*". Part of such approach or direction leads to understanding the vulnerability of communities, the environment, and natural resources that have been used, misused, and abused for a long.

Clearly, from what we experience during the COVID-19 pandemic, resource efficiency and distribution are areas that require more attention. More precisely, from the environmental or natural resource management, we have seen many flaws that should be assessed carefully. From optimal resource planning (Petrović 2020) to sustainable resource management (You et al. 2020), we see a wide range of weaknesses and opportunities to see better use/enhancement of the environments and accessibility to them protecting them, etc. As argued by Rousseau and Deschacht (2020), during the COVID-19, "*the increased experience with local natural resources may strengthen public support for a recovery program that puts the transition towards a more sustainable economic system centrally*". Hence, we see a positive shift in people's appreciation of the natural environments and resources, highlighting their vulnerabilities and lack of adequate governance in general.

Cities continuously face a variety of shortages, including natural resources. From water and food supplies to other provisions and amenities (Cheshmehzangi 2020d), natural resource shortages could cause conflicts and social unrest. Such situations put pressure on sustainable peace in cities and communities where natural environments and resources are scarce or only available to society's elite members. In these areas, we also note the importance of resilience during the pandemic (ibid), showing a range of productivity issues and resource limitations (Ratner et al. 2020). From the natural resource management perspective, we also see growing concerns on how solidarity and decision-making were maintained (Walters et al. 2021). An example of these would be abundant natural resources, which are investigated as a significant part of boosting economic growth (Ridzuan et al. 2021). Thus, we see more opportunities arising, which show resource limitations are related to some of the essential resources and natural resources that are yet to be studied in future research.

Let us consider the COVID-19 as a natural disaster (Seddighi 2020) rather than a health emergency (Cheshmehzangi 2020d). We see large-scale deficiencies in resource efficiency and distribution, particularly from the environmental/natural perspectives. Lack of access to clean water, adequate dietary and food, and lack of clean living environments in cities are of concern to many countries that struggle to safeguard peace and well-being in their societies. Cities are playful living environments but commonly lack a harmonious balance between their natural and built environments. Therefore, we see growing attention on the availability and accessibility factors, ensuring that natural resources are part of the living environments for healthy communities.

8.3 Addressing the Flaws

Natural resource availability is a context-specific matter, but accessibility to them is a matter of choice made by decision-makers and those that could sustain peace in cities and communities. During the COVID-19 pandemic, we see a wide range of deficiencies related to natural resources and their equitable distributions. Set aside the complexity of resource availability for other factors, such as healthcare resources and essential supplies, we see the role of natural resources has become even more critical. Examples of having the right to access natural environments are just a simple factor representing inequalities between different demographics and social classes. From accessibility to private and public green and natural environments to the distribution of natural resources such as clean water and food, we see significant disparities between those at the top and the bottom of the pyramid.

In addition, we see environmental or natural governance as a significant driver to how peace and health could be achieved and/or maintained in cities. Despite the fact natural environments are often a small proportionate of our cities, they have significant impacts on healthy living styles and achieving healthy cities. The correlation between environmental, social, and governance (ESG) harmonisation is not just part of the materiality of discussing sustainability (Adams and Abhayawansa 2021), but also associated with resource efficiency, accessibility, and fair distributions to all. To address some of the existing flaws, we have to explore the issues that have, for long, become the normality of grassroots and vulnerable groups of society. We then need to make choices rather than mere top-down decisions (Cheshmehzangi and Dawodu 2019; Cheshmehzangi et al. 2021). The availability of natural resources should be prioritised, as they have also been central to several SDGs. Moreover, the approach to enhancing accessibility and providing equality in the public would enable us to achieve better public health focused on well-being and fairness in society. In doing so, we are able to maintain peace in communities, ensuring resource availability and their equitable distributions are not just limited to natural resources. While we focused on environmental/natural resources in this chapter, we cannot neglect the importance of other resources during the pandemics. More importantly, resource limitations of any sort could also cause social unrest and potential conflicts between communities. In this regard, we see a growing gap that defines the mounting inequalities of more impoverished communities.

As a response to the COVID-19 pandemic, resource management was considered a critical aspect of enhancing urban resilience (Cheshmehzangi 2020d). More to this is the fairness in ensuring resource efficiency and distribution, particularly from the environmental sustainability perspective. If we aim for healthy people and cities, we have to ensure resources are made available to wider social groups. And if we strive to maintain peace in such healthy communities, we cannot divorce environmental governance from general governance. Thus, in achieving good governance, people's role and involvement are a necessity, or else we may fail in safeguarding health and peace in communities.

8.4 A Summary

In this chapter, we stressed three areas of natural resource governance, natural governance, and environmental governance. While different, the three are interlinked and somehow complemented each other. In the face of the rising ESG, as Diaz et al. (2021) argue, we ought to reconsider systematic factors and aim to evaluate and share our resources more cautiously—but considerately. However, from the environmental/natural dimension, we are still far behind in achieving the ideals of SDGs. Directions, such as natural resource management, are essential to maintain peace and achieve healthy communities. Such mentality could help us improve several overarching aspects, such as environmental regulations, natural resources' availability and accessibility, and environmental externalities. Our environmental awareness should certainly move towards higher-level social responsibility (Severo et al. 2021), allowing for better environmental management strategy (Kang et al. 2021), the gradual reduction of inequalities and vulnerabilities, and urban resilience enhancement.

As we close the environmental part of the book, we verify the general lack of attention to environmental/natural aspects, particularly from the HSP perspective. While environmental aspects are related to other dimensions of economic, social, and institutional, we note they themselves are important as separate aspects, too. In this part, we tried to cover areas that are often studied differently in peace–sustainability studies, ensuring that we do not neglect the role of health and well-being in safeguarding the environmental sustainability of cities and communities. In the next part, we move into the economic dimension, which again would include several key aspects that require our attention in maintaining a suitable HSP nexus in cities.

References

Adams CA, Abhayawansa S (2021) Connecting the COVID-19 pandemic, environmental, social and governance (ESG) investing and calls for 'harmonisation' of sustainability reporting. Critical Perspectives on Accounting. 102309. https://doi.org/10.1016/j.cpa.2021.102309

Agbali M, Trillo C, Fernando TP, Arayici Y (2017) Creating smart and healthy cities by exploring the potentials of emerging technologies and social innovation for urban efficiency: lessons from the innovative city of Boston. Int J Energ Power Eng 11(5)

Bajaj M, Chiu B (2009) Education for sustainable development as peace education. Peace Change 34 (4):441–455

Brauch HG, Spring ÚO, Grin J, Scheffran J (eds) (2016) Handbook on sustainability transition and sustainable peace, vol 10. Springer, Cham

Bretthauer JM (2015) Conditions for peace and conflict: applying a fuzzy-set qualitative comparative analysis to cases of resource scarcity. J Conflict Resolut 59 (4):593–616

Brock L (1991) Peace through parks: the environment on the peace research agenda. J Peace Res 28(4):407–423

Buckles D (ed) (1999) Cultivating peace: conflict and collaboration in natural resource management. International development research centre (IDRC), Canada, world bank institute, international bank for reconstruction and development staff

Cheshmehzangi A (2020a) Resilience first briefing—urban resilience around COVID-19: lessons from China, presented in 'Resilience first briefing', held in London, UK, pp 1–13

Cheshmehzangi A (2020b) Introduction: the city during outbreak events. In: Cheshmehzangi A (ed) The city in need: urban resilience and city management in disruptive disease outbreak events. Singapore: Springer. pp 1–15

Cheshmehzangi A (2020c) Housing and health evaluation related to general comfort and indoor thermal comfort satisfaction during the COVID-19 lockdown. J Hum Behav Soc Environ 31(1–4):184–209

Cheshmehzangi A (2020d) The city in need: urban resilience and city management in disruptive disease outbreak events. Springer, Singapore

Cheshmehzangi A, Dawodu A (2019) The review of sustainable development goals (sdgs): people, perspective and planning. In: Cheshmehzangi A, Dawodu A (eds) Sustainable urban development in the age of climate change—people: The Cure or Curse. Palgrave Macmillan, Singapore, pp 133–156

Cheshmehzangi A, Dawodu A, Sharifi A (2021) Sustainable urbanism in China. Routledge, New York and London

Corburn J (2009) Toward the healthy city: people, places, and the politics of urban planning. MIT Press, Cambridge, MA

Davies-Vengoechea X (2004) A positive concept of peace. Keeping the peace: conflict resolution and peaceful societies around the world. pp 11–18

DeWit A (2020) Resource-efficiency and critical raw materials in Covid-19 and recovery. Rikkyo Econ Rev. 74(1):1–58. Available from: https://economics.rikkyo.ac.jp/research/paper/pudcar00000002ed-att/p001-058_74-1.pdf

Di Maio F, Rem PC, Baldé K, Polder M (2017) Measuring resource efficiency and circular economy: a market value approach. Resour Conserv Recycl 122:163–171

Diaz V, Ibrushi D, Zhao J (2021) Reconsidering systematic factors during the COVID-19 pandemic—the rising importance of ESG. Finan Res Lett 38:101870

Duhl LJ, Sanchez AK, World Health Organization (1999) Healthy cities and the city planning process: a background document on links between health and urban planning (No. EUR/ICP/CHDV 03 04 03). Copenhagen: WHO Regional Office for Europe

Ekins P, Hughes N, Brigenzu S, Arden Clark C, Fischer-Kowalski M, Graedel T, Hajer M, Hashimoto S, Hatfield-Dodds S, Havlik P, Hertwich E (2016) Resource efficiency: potential and economic implications. Report of the international resource panel, united nations environment program (UNEP), Paris. Available from: http://pure.iiasa.ac.at/id/eprint/13813/

Freire-González J, Vivanco DF (2020) Pandemics and the environmental rebound effect: reflections from COVID-19. Environ Resour Econ 1–4

Gutiérrez-Montes IA (2005) Healthy communities equal healthy ecosystems? Evolution (and breakdown) of a participatory ecological research project towards a community natural resource management process. San Miguel Chimalapa (Mexico). Available from: https://lib.dr.iastate.edu/rtd/1735/

Happaerts S (2016) Discourse and practice of transitions in international policymaking on resource efficiency in the EU. In: Brauch HG, Spring UO, Grin J, Scheffran J (eds) Handbook on sustainability transition and sustainable peace. Springer, Cham, pp 869–884

Harris IM (2004) Peace education theory. J Peace Educ 1 (1):5–20

Helm D (2020) The environmental impacts of the coronavirus. Environ Resource Econ 76:21–38

Kang A, Ren L, Hua C, Song H, Dong M, Fang Z, Zhu M (2021) Environmental management strategy in response to COVID-19 in China: based on text mining of government open information. Sci Total Environ 769:145158

Krampe F (2017) Toward sustainable peace: a new research agenda for post-conflict natural resource management. Global Environ Polit 17(4):1–8

Marton-Lafevre J (2007) Peace parks: conservation and conflict resolution. MIT Press, Cambridge, MA
Mori H, Takahashi Y, Zusman E, Mader A, Kawazu E, Otsuka T, Moinuddin M, King P, Elder M, Teoh WCS, Takeda T (2020) Implications of COVID-19 for the environment and sustainability. Available from: https://www.iges.or.jp/en/pub/covid19-e/en
Morris D (1987) Healthy cities: self-reliant cities. Health Promot Int 2(2):169–176
Nellemann C, Henriksen R, Kreilhuber A, Stewart D, Kotsovou M, Raxter P, Mrema E, Barrat S (2016) The rise of environmental crime: a growing threat to natural resources, peace, development and security. United Nations Environment Programme (UNEP)
Petrović N (Sep 2020) Simulation environment for optimal resource planning during COVID-19 crisis. In: 2020 55th international scientific conference on information, communication and energy systems and technologies (ICEST). IEEE, pp 23–26
Ramaswami A, Russell AG, Culligan PJ, Sharma KR, Kumar E (2016) Meta-principles for developing smart, sustainable, and healthy cities. Science 352 (6288):940–943
Ratner, L., Martin-Blais, R., Warrell, C., & Narla, N. P. (2020). Reflections on Resilience during the COVID-19 Pandemic: Six Lessons from Working in Resource-Denied Settings. The American journal of tropical medicine and hygiene, 102(6), 1178-1180
Ridzuan AR, Shaari MS, Rosli A, Jamil ARM, Siswantini S, Lestari A, Zakaria S (2021) The nexus between economic growth and natural resource abundance in selected ASEAN countries before pandemic Covid-19. Int J Energ Econ Policy 11(2):281
Rissler GE, Shields PM (2019) Positive peace—A necessary touchstone for public administration.Adm Theory Prax, 41(1):60–78
Rousseau S, Deschacht N (2020) Public awareness of nature and the environment during the COVID-19 crisis. Environ Resour Econ 76(4):1149–1159
Sarkar P, Debnath N, Reang D (2021) Coupled human-environment system amid COVID-19 crisis: a conceptual model to understand the nexus. Sci Total Environ 753:141757
Schafer J, Black R (2003) Conflict, peace, and the history of natural resource management in Sussundenga district, Mozambique. Afr Stud Rev 55–81
Seddighi H (2020) COVID-19 as a natural disaster: focusing on exposure and vulnerability for response. Disaster Med Public Health Prep 14(4):e42–e43
Severo EA, De Guimarães JCF, Dellarmelin ML (2021) Impact of the COVID-19 pandemic on environmental awareness, sustainable consumption and social responsibility: evidence from generations in Brazil and Portugal. J Cleaner Prod 286:124947
Shields PM, Soeters J (2017) Peaceweaving: Jane Addams, positive peace, and public administration. Am Rev Public Adm 47(3):323–339
Walters G, Broome N, Cracco M, Dash T, Dudley N, Elías S, Hymas O, Mangubhai S, Mohan V, Niederberger T, Achtone C (2021) COVID-19, indigenous peoples, local communities and natural resource governance. Parks 27:57–62
You S, Sonne C, Ok YS (2020) COVID-19's unsustainable waste management. Science 368(6498):1438
Zhang Y, Yang Z (2007) Eco-efficiency of urban material metabolism: a case study in Shenzhen, China. Acta Ecol Sin 27(8):3124–3131
Zhou X (2006) The Canada healthy city program: practice and experience [J]. J Public Manag 3. https://en.cnki.com.cn/Article_en/CJFDTotal-GGGL200603013.htm

Part III
Sustainability and Peace: "Urban Health from the Economic Dimension"

In the third part of studying sustainability and peace, we present three primary aspects representing urban health from the economic dimension.

The first aspect is a healthy or sound business environment, which is also one of the main pillars of positive peace. From the economic perspective, this is a significant factor during the disruptive times of pandemics. The adversities that affect the operation of businesses could destabilise the city's overall operations. These points are discussed in Chap. 9, focused on maintaining a healthy business environment in cities and communities. The second aspect of this part is a low level of corruption, a crossover between economy and institutional structures. Again, this is one of the main pillars of positive peace, which is very important for economic growth, stability, and resilience. Our arguments here would provide some of the issues that cities faced during the COVID-19 pandemic. Moreover, there would be some crossover discussions with other dimensions. This aspect is studied in Chap. 10, where we aim to explore the importance of low-level corruption in sustaining health and peace in cities. The third aspect looks at one specific factor of the supply chain. As we noted (and experienced) during the COVID-19 pandemic, our supply chain could be vulnerable, particularly in processes that move us away from self-sufficient economies and development. This important aspect is presented in Chap. 11, identifying the role of the supply chain in achieving urban health, sustainability, and peace.

9 Healthy Business Environment and Facing Intolerable Disruptions

The greatest ones grow up while growing their visions.

—Ali Cheshmehzangi, 2018

9.1 A Brief Exploration: Healthy or Sound Business Environment in Cities

As Forrer et al. (2012) suggest, there are ways that businesses can foster peace, ideally with specific "*policies, characteristics, and operating environments*" for peace-promoting behaviour. However, to have businesses engaged in such processes (Saaty and Vargas 2013; Forrer and Katsos 2015), we ought to consider the value of a sound business environment in cities. This reflects on earlier work by Rosenberg (1996), where we see a closer relationship between peace, health, and the environment. While her focus was on the business environment per se, but several health factors were discussed as part of peacebuilding processes. During the COVID-19 pandemic, we note several issues on business risks and the well-being of the business environment. From the business-as-usual perspective, we realise problems related to continuous disruptions on sustainable global peace (Amadei 2021), highlighting unsustainable cycles of economic development and never-ending growth (Cheshmehzangi 2020a). In this regard, the well-being of businesses and economic development would undoubtedly be related to the actual health of the business environment. Nevertheless, some problems exist, meaning that we may need to move towards new normals for a sound business environment (Burkett and Moore 2020) or aim to have more interventions to keep our economies healthy and resilient.

Many aspects related to a healthy business environment lead to potential business model innovation (Breier et al. 2021) or innovative technologies (Di Vaio et al. 2020), highlighting the COVID-19 pandemic. On the other hand, with many business failures, we see the importance of economic resilience and particularly towards peacebuilding practices (Schroff 2020). Some studies look at benefit–cost analysis of economic resilience to evaluate and model the economic impact of COVID-19 (Rose and Wei 2020), while other studies bring in other aspects that determine economic difficulties in specific contexts and/or sectors (Kimhi et al. 2020). There is, however, a growing argument around diversification and growth strategies (World Bank 2020a, b), which indicate not only approaches towards peace operations (de Coning

A. Cheshmehzangi, *Urban Health, Sustainability, and Peace in the Day the World Stopped*, Sustainable Development Goals Series, https://doi.org/10.1007/978-981-16-4888-5_9

2020), but also new opportunities for resilience enhancement (Cheshmehzangi 2020b). Hence, we argue motives for sustainable economic strategies, such as those that would entail diversification, building resilience, and peacebuilding.

As an entry point to the economic dimension, this chapter serves as a starting point on the healthiness and well-being of the business environments. We delve into some sector-specific examples with the aim to highlight issues of peace and security in relation to the economic sustainability of cities and communities. In general, we aim to learn from economic resilience strategies that aim to oppose globalisation and market governance issues (Jenny 2020) and realise what makes the cycle of short-term benefits and future resilience (Gössling 2020) pathways towards sustainability and stability. Partly, we must understand and measure economic risks on businesses (Cheshmehzangi et al. 2020b; Noy et al. 2020) and evaluate specific issues or patterns that may lead to business fragility and failures. Some argue in favour of the empowerment strategies (Berawi 2020), and some suggest the need to "*provide additional economic resilience*" (Balanton et al. 2020). In both scenarios, we ought to consider a sort of equilibrium that seems to be missing in contexts where disparities are widespread, poverty is increasing, and inequalities are embedded in their daily operations.

While we hope to cover much of the aspects related to the health–sustainability–peace (HSP) nexus, we can only cover some of the key concerning areas regarding the healthiness of our business environments. In particular, we aim to cover areas of business management (Sheth 2020) and particular implications (Craven et al. 2020) that led to significant economic failures. This chapter covers these topics from the economic sustainability perspective. It also highlights common adversities that affected the operation of businesses, some that led to destabilisation of city operations, and some that helped us understand that the previous normal may not have been the best operational option.

9.2 Reflecting on How Cities Reacted During the Pandemic

During the COVID-19 pandemic, the most considerable impact has been on small and medium-sized enterprises (SMEs) (Cheshmehzangi 2020b) or just small businesses (Baker and Judge 2020; Bartik et al. 2020). While some studies find interesting results in terms of business performances and expectations and how they reacted to pandemic crises (Verma and Gustafsson 2020; Cirera et al. 2021; Meyer et al. 2021), we also note the changing behaviours across all operations (Cheshmehzangi 2020c). In this regard, we see the pandemic crisis has not been equal across the board, affecting the most those vulnerable businesses of smaller scale, such as local businesses. This also indicates the problematic situations of the capitalist systems, particularly those set in favour of larger-scale enterprises and businesses. The situation has for long created an unhealthy business environment that lacks equality for SMEs and local businesses. Furthermore, the examples of SME failures (Gourinchas et al. 2020) and inadequate policy responses to respond to COVID-19 impact on them (Cepel et al. 2020; Cowling et al. 2020; Juergensen et al. 2020) show few options for business resilience and more about business survival (Fitriasari 2020). Some of the primary issues include liquidity needs (McGeever et al. 2020), economic recovery plans (Yu 2020), and complexities that still exist in the business model operations (Priyono et al. 2020). In this regard, we see the scale of economic impact on businesses could differ vastly from context to context and sector to sector.

It is indeed very unfortunate to see our strategies are often focused on survival strategies (Omar et al. 2020) than enhancing the resilience and healthiness of our businesses and their environments. An obvious example is the wide range of impacts on the tourism and hospitality businesses (Cheval et al. 2020), which shows a lack of control and deficient crisis management practices. The fragile system that shapes the tourism and hospitality industry (Kaushal and Srivastava 2021) undermines other primary aspects such as human rights (Baum and Hai 2020), resilience strategies against unprecedented challenges (Gursoy and Chi 2020), and intervention policies (Huang et al. 2020). We note that the lack of response, protocol, and management is not related to businesses necessarily, but their environments. The financial distress among many related firms (Crespí-Cladera et al. 2021) highlights the vulnerability of industries that very much depend on unsustainable cycles of expansion and growth. Issues related to movement control (Omar et al. 2020), preparedness plan, and reflective response (Cheshmehzangi 2020b) have caused severe impacts on the sustainability of operations and economic stability, leading to unhealthy situations that could trigger unrests against peace in communities.

Studies that explore macroeconomic issues identify very clearly how real and financial social accounting matrix (FSAM) is essential in evaluating the impacts of the pandemic on businesses such as SMEs (Pedauga et al. 2021). The so-called COVID-19 crisis has already affected the performance of many businesses (Albonico et al. 2020). Despite the fact some of the solutions involve technology adoption (Kumar and Ayedee 2021) and emerging state-of-the-art technologies (Akpan et al. 2020) in business environments, we note deficiencies in developing or sustaining resilience across the board. The lack of sustainable business models (Gregurec et al. 2021) shows that "*businesses have been exposed to various challenges during the global pandemic, and their response to this disruption has impacted their resilience as well as their chances to overcome this crisis*". Thus, we see continuing effects and potentially a much longer impact on the economic sustainability and peacebuilding processes after the pandemic is over. Despite growing vulnerabilities and intensifying poverty, we see some governmental initiatives favouring increasing real estate prices, infecting the market (Del Giudice et al. 2020; Marona and Tomal 2020), leading to unhealthy workflows, higher debt levels, and an eventual stock market crash.

Yet again, we are in a position knowing long-term adversities disrupting the business environments, especially as the current uncertain economic situation (Worzala 2020) that may cause more turbulence in future policies, stock market reaction (Baker et al. 2020), and cost–benefit analysis of specific businesses (Rowthorn and Maciejowski 2020). From the inception of the COVID-19 pandemic, we note global issues that suggest fragile international cooperation (Brown and Susskind 2020). The problems of supply and demand stocks during the COVID-19 pandemic (del Rio-Chanona et al. 2020) show the unhealthy global trade markets that have for long structured as our globalised operations of businesses. The situation, as it persists, highlights a wide range of uncertainties, undermining peace in economic sustainability. During the COVID-19, we note that the fat cats of society earned the most, affecting the operations as a whole. The rich got richer while causing a wider distribution of the pandemic at first. The poor became more vulnerable as their daily needs led to more inequalities and widening gaps between social classes. While the role of business model innovation (Brier et al. 2021) became more apparent, we see growing issues in established business models, management of business environments, and sustaining peace in economic stability.

9.3 Addressing the Flaws

At any time in human history, peace in a business environment is vital to peace in society. We learnt that situations like health crises and emergencies intensify economic crises (Borio 2020), creating a sort of imbalance between businesses, operations, and peace. Issues of capitalism were highlighted as part of the crisis (Murshed 2020), leading to a better understanding of cross-country, sub-regional, and international trade issues, widening disparities, and unequal mindsets. The values of peace and quality, as we usually know them, suggest to us the importance of reorganising the current business environments, the transformation of practices, and restricting the operations (Cheshmehzangi 2020b; Lew et al. 2020). By debating between a new normal or business-as-usual (BAU), some scholars, like Yap (2020), suggest lessons that could be learnt from the financial crisis perspective, enabling us to reconsider the well-being of our business environments. Thus, we can verify structural issues that have rotten the backbone of business operations, corrupting the business environment that is no longer equitable and equal for the majority.

To respond to some of these common flaws, governments should revisit the unsustainable patterns of economic expansion for larger businesses, not knowing that their never-ending long-term trends of development could widen inequalities in the business environment. To ensure businesses are safe and resilient, we ought to make their environments healthy and resilient. In order to do so, we need to meet the bottom-up demand and reconsider the support and services that have been decaying for a long. The corrupted business environment led by the elites creates uncertain sociopolitical situations that may result in mismanagement of businesses and their environments against any sort of crisis. The COVID-19 pandemic puts pressure on the peace and security agenda (Warren and Bordoloi 2020) not because of its impacts but because our business and economic systems are ill within. The infected business environments have been ignored for a long by governments who themselves are direct beneficiaries of monopoly-based business environments. Such issues have led to corrupted environments that limit the growth of smaller businesses, put pressures on those with little profit, and create unhealthy situations for societies.

9.4 A Summary

If we reflect more carefully, we see the fat cats are fatter, and the poor are skinnier. This is the result of unhealthy business environments, which by all means, cannot encourage peacebuilding. Instead, such a situation keeps the fundamental values in mind, which are generally appreciated as unhealthy productions and general consumption models. Such behaviours and mindsets have caused us a fortune, leading to corrupted business environments, i.e. health and well-being are for those on top of the pyramid and not those who suffer from long-lasting inequalities. To ensure peaceful business environments, we should also ensure fair distribution of equity among all business sectors and sizes. In doing so, regime types may need to be re-evaluated, and our economic sustainability should be refined.

To reflect more explicitly on the impacts of the COVID-19 pandemic, we note many uncertainties of the aftermath for businesses, particularly in the global south (Nseobot et al. 2020). We report the importance of interventions and innovative strategies against the current BAUs. Also, we note deficiencies of the global trade and market, suggesting larger-scale impacts on the global economy (Mishra 2020) due to the pandemic crisis. To summarise briefly, we looked into some evident and non-obvious impacts of the pandemic on businesses and their operations. More explicitly, in this chapter, we highlighted the importance of the business environment in achieving the health–sustainability–peace (HSP) nexus for businesses. This is indeed very critical to what will be discussed in the following two chapters.

References

Akpan IJ, Udoh EAP, Adebisi B (2020) Small business awareness and adoption of state-of-the-art technologies in emerging and developing markets, and lessons from the COVID-19 pandemic. J Small Bus Entrepreneurship 1–18

Albonico M, Mladenov Z, Sharma R (2020) How the COVID-19 crisis is affecting UK small and medium-size enterprises. McKinsey & Company. 5p document, Available from: https://www.mckinsey.com/~/media/McKinsey/Industries/Public%20and%20Social%20Sector/Our%20Insights/How%20the%20COVID%2019%20crisis%20is%20affecting%20UK%20small%20and%20medium%20size%20enterprises/How-the-COVID-19-crisis-is-affecting-UK-small-and-medium-size%20enterprises.pdf

Amadei B (2021) Systemic modeling of the peace-development nexus. Sustainability 13(5):2522

Baker SR, Bloom N, Davis SJ, Kost K, Sammon M, Viratyosin T (2020) The unprecedented stock market reaction to COVID-19. Rev Asset Pricing Stud 10 (4):742–758

Baker T, Judge K (2020) How to help small businesses survive COVID-19. Columbia law and economics working paper, (620)

Bartik AW, Bertrand M, Cullen ZB, Glaeser EL, Luca M, Stanton CT (2020) How are small businesses adjusting to COVID-19? Early evidence from a survey (No. w26989). Natl Bur Econ Resear

Baum T, Hai NTT (2020) Hospitality, tourism, human rights and the impact of COVID-19. Int J Contemp Hospitality Manag Available from: https://doi.org/10.1108/IJCHM-03-2020-0242/

Berawi MA (2020) Empowering healthcare, economic, and social resilience during global pandemic COVID-19. Int J Technol 11(3):436–439

Blanton RE, Mock NB, Hiruy HN, Schieffelin JS, Doumbia S, Happi C, Samuels RJ, Oberhelman RA (2020) African resources and the promise of resilience against COVID-19. Am J Trop Med Hyg 103(2):539

Borio C (2020) The Covid-19 economic crisis: dangerously unique. Bus Econ 55(4):181–190

Breier M, Kallmuenzer A, Clauss T, Gast J, Kraus S, Tiberius V (2021) The role of business model innovation in the hospitality industry during the COVID-19 crisis. Int J Hospitality Manag 92:102723

Brown G, Susskind D (2020) International cooperation during the COVID-19 pandemic. Oxford Rev Econ Policy 36(Supplement_1):S64–S76

Burkett M, Moore N (2020) This is just the beginning: climate change, positive peace, and the "New normal". Honolulu: University of Hawai'i Press and the Center for Biographical Research. Available from: https://library.oapen.org/bitstream/handle/20.500.12657/43421/external_content.pdf?sequence=1#page=36

Cepel M, Gavurova B, Dvorský J, Belas J (2020) The impact of the COVID-19 crisis on the perception of business risk in the SME segment. J Int Stud. Available from: https://publikace.k.utb.cz/handle/10563/1009961

Cheshmehzangi A (2020b) The city in need: urban resilience and city management in disruptive disease outbreak events. Springer, Singapore

Cheshmehzangi A (2020c) COVID-19 and household energy implications: what are the main impacts on energy use? Heliyon 6(10):e05202. https://doi.org/10.1016/j.heliyon.2020.e05202

Cheshmehzangi A (2020a) Recommendations for 'The city in need.' In: Cheshmehzangi A (ed) The city in need: urban resilience and city management in disruptive disease outbreak events. Springer, Singapore, pp 285–304

Cheval S, Mihai Adamescu C, Georgiadis T, Herrnegger M, Piticar A, Legates DR (2020) Observed and potential impacts of the COVID-19 pandemic on the environment. Int J Environ Resear Public Health 17(11):4140

Cirera X, Cruz M, Davies E, Grover A, Iacovone L, Cordova JEL, Medvedev D, Maduko FO, Nayyar G, Reyes Ortega S, Torres J (2021) Policies to support businesses through the COVID-19 shock: a firm level perspective. World Bank Resear Observer 36(1):41–66

Cowling M, Brown R, Rocha A (2020) <? covid19?> Did you save some cash for a rainy COVID-19 day? The crisis and SMEs. Int Small Bus J 38(7):593–604

Craven M, Liu L, Mysore M, Wilson M (2020) COVID-19: implications for business. McKinsey Company 1–8

Crespí-Cladera R, Martín-Oliver A, Pascual-Fuster B (2021) Financial distress in the hospitality industry during the Covid-19 disaster. Tourism Manag 85:104301

de Coning C (2020) The impact of COVID-19 on peace operations in Africa. Conflict and resilience monitor. Available from: https://www.accord.org.za/conflict-trends/the-impact-of-covid-19-on-peace-operations-in-africa/

Del Giudice V, De Paola P, Del Giudice FP (2020) COVID-19 infects real estate markets: short and mid-run effects on housing prices in Campania region (Italy). Soc Sci 9(7):114

del Rio-Chanona RM, Mealy P, Pichler A, Lafond F, Farmer JD (2020) Supply and demand shocks in the COVID-19 pandemic: an industry and occupation perspective. Oxford Rev Econ Policy 36(Supplement_1):S94–S137

Di Vaio A, Boccia F, Landriani L, Palladino R (2020) Artificial intelligence in the agri-food system: rethinking sustainable business models in the COVID-19 scenario. Sustainability 12(12):4851

Fitriasari F (2020) How do small and medium enterprise (SME) survive the COVID-19 outbreak?. Jurnal Inovasi Ekon 5(02). Available from: http://202.52.52.22/index.php/JIKO/article/view/11838

Forrer JJ, Fort TL, Gilpin R (2012) How business can foster peace. US Institute of Peace. Available from: https://www.jstor.org/stable/pdf/resrep12185.pdf

Forrer JJ, Katsos JE (2015) Business and peace in the buffer condition. Acad Manag Perspect 29(4):438–450

Gössling S (2020) Risks, resilience, and pathways to sustainable aviation: a COVID-19 perspective. J Air Transp Manage 89:101933

Gourinchas PO, Kalemli-Özcan Ş, Penciakova V, Sander N (2020) Covid-19 and SME failures (No. w27877). Natl Bur Econ Resear

Gregurec I, Tomičić Furjan M, Tomičić-Pupek K (2021) The impact of COVID-19 on sustainable business models in SMEs. Sustainability 13(3):1098

Gursoy D, Chi CG (2020) Effects of COVID-19 pandemic on hospitality industry: review of the current situations and a research agenda. J Hosp Market Manag 29(5):527–529

Huang A, Makridis C, Baker M, Medeiros M, Guo Z (2020) Understanding the impact of COVID-19 intervention policies on the hospitality labor market. Int J Hospitality Manag 91:102660

Jenny F (Mar 28, 2020) Economic resilience, globalization and market governance: facing the Covid-19 test. Globalization and market governance: facing the COVID-19 test

Juergensen J, Guimón J, Narula R (2020) European SMEs amidst the COVID-19 crisis: assessing impact and policy responses. J Ind Bus Econ 47(3):499–510
Kaushal V, Srivastava S (2021) Hospitality and tourism industry amid COVID-19 pandemic: perspectives on challenges and learnings from India. Int J Hospitality Manag 92:102707
Kimhi S, Marciano H, Eshel Y, Adini B (2020) Recovery from the COVID-19 pandemic: distress and resilience. Int J Disaster Risk Reduction 50:101843
Kumar A, Ayedee D (2021) Technology adoption: a solution for SMEs to overcome problems during COVID-19. Forthcoming, Acad Mark Stud J 25(1)
Lew AA, Cheer JM, Haywood M, Brouder P, Salazar NB (2020) Visions of travel and tourism after the global COVID-19 transformation of 2020. Tour Geogr 22 (3):455–466
Lu Y, Wu J, Peng J, Lu L (2020) The perceived impact of the Covid-19 epidemic: evidence from a sample of 4807 SMEs in Sichuan province China. Environ Hazards 19(4):323–340
Marona B, Tomal M (2020) The COVID-19 pandemic impact upon housing brokers' workflow and their clients' attitude: real estate market in Krakow. Entrepreneurial Bus Econ Rev 8(4):221–232
McGeever N, McQuinn J, Myers S (2020) SME liquidity needs during the COVID-19 shock (No. 2/FS/20). Central Bank of Ireland
Meyer BH, Prescott B, Sheng XS (2021) The impact of the COVID-19 pandemic on business expectations. Int J Forecast. In press. Available from: https://doi.org/10.1016/j.ijforecast.2021.02.009
Mishra MK (2020) The world after COVID-19 and its impact on global economy. Available from: https://www.econstor.eu/handle/10419/215931
Murshed SM (2020) Capitalism and COVID-19: crisis at the crossroads. Peace Econ, Peace Sci Public Policy 26(3)
Noy I, Doan N, Ferrarini B, Park D (2020) Measuring the economic risk of COVID-19. Global Pol 11(4):413–423
Nseobot IR, Simeon II, Effiong AI, Frank EI, Ukpong ES, Essien MO (2020) COVID-19: the aftermath for businesses in developing countries. Int J Bus Educ Manag Stud (IJBEMS). Available from: https://papers.ssrn.com/sol3/papers.cfm?abstract_id=3592603
Omar ARC, Ishak S, Jusoh MA (2020) The impact of Covid-19 movement control order on SMEs' businesses and survival strategies. Geografia-Malays J Soc Space 16(2)
Pedauga L, Sáez, F., & Delgado-Márquez, B. L. (2021). Macroeconomic lockdown and SMEs: the impact of the COVID-19 pandemic in Spain. Small Bus Econ 1–24
Priyono A, Moin A, Putri VNAO (2020) Identifying digital transformation paths in the business model of SMEs during the COVID-19 pandemic. J Open Innov: Technol, Mark, Complex 6(4):104
Rose A, Wei D (2020) Modeling the economic impact of COVID-19. Available from: https://www.remi.com/wp-content/uploads/2020/03/USC_REMI_Webinar_COVID-19_Rose-and-Wei_3-18-20F.pdf
Rosenberg DG (1996) Peace, health and the environment: challenging a conspiracy of silence. Peace Res 28 (1):93–94
Rowthorn R, Maciejowski J (2020) A cost–benefit analysis of the COVID-19 disease. Oxford Rev Econ Policy, 36(Supplement_1):S38–S55
Saaty TL, Vargas LG (2013) The logic of priorities: applications of business, energy, health and transportation. Springer Sci Bus Media
Schroff J (2020) International business for peace: adopting an academic peace education approach to encourage corporate peacebuilding. Available from: repositori.uji.es/xmlui/handle/10234/192352
Sheth J (2020) Business of business is more than business: managing during the Covid crisis. Ind Mark Manage 88:261–264
Verma S, Gustafsson A (2020) Investigating the emerging COVID-19 research trends in the field of business and management: a bibliometric analysis approach. J Bus Res 118:253–261
Warren MA, Bordoloi S (2020) When COVID-19 exacerbates inequities: the path forward for generating wellbeing. Int J Wellbeing 10(3)
World Bank (2020a) Central African republic economic update, October 2020: the central African republic in times of COVID-19-diversifying the economy to build resilience and foster growth. Available from: https://doi.org/10.1596/34803
World Bank (2020b) Breaking out of fragility: a country economic memorandum for diversification and growth in Iraq. Available from: https://doi.org/10.1596/978-1-4648-1637-6
Worzala E (2020) COVID 19, real estate and uncertainty: examining this new "normal" through the quotes of Jim Graaskamp. J Property Investment Finan
Yap OF (2020) A new normal or business-as-usual? Lessons for COVID-19 from financial crises in East and Southeast Asia. Eur J Dev Resear 32(5):1504–1534

10 Low Level of Corruption: A Crossover Between Economy and Institutional Structures

Corruption and peace are two sides of humanity's spectrum.

—Ali Cheshmehzangi, 2021

10.1 A Brief Exploration: Low Level of Corruption in Achieving Economic Sustainability

In general, peace and sustainable security are the backbones of anticorruption strategies and towards sustainable development (Newham 2014). As suggested by Khemani (2020), the COVID-19 pandemic is an excellent opportunity to build legitimacy and trust in public institutions, helping to expose corrupt politicians and move towards a low level of corruption. Many studies relate a low level of corruption with peacebuilding, suggesting the correlation between a low level of corruption and a high(er) level of democracy (McMann et al. 2019), or the relationship between sociopolitical and economic conditions and low-level corruption (Chinweuba 2020). In all cases, the author notes the central position of low-level corruption, favouring peace and peacebuilding processes. Thus, achieving sustainable development would really depend on lowering the corruption levels, allowing for better relationships, paradigm shifts, and longer-term solutions against unplanned and unstructured conditions.

As highlighted by Jain (2001), the political economy of corruption is complicated, which becomes ever more complicated in peacebuilding procedures and post-conflict situations (Le Billon 2008; Rose-Ackerman 2008). We note a significant part of it is related to governance aspects, but a low level of corruption could also be embedded in the economic development and economic sustainability of cities and communities. Examples such as power sharing in the political economy structures (Haass and Ottmann 2017) and economic functions in society (Le Billon 2003) highlight the importance of low-level corruption in achieving sustainable economies. Hence, low corruption levels play a central role in maintaining a healthy economy, especially when dealing with calamities, emergencies, and similar events. In a way, low-level corruption means that institutional structures are healthy, and their operations are reasonably decent. However, in the case of higher-level corruption, we expect to see situations that suggest a non-operational status of institutions, their

A. Cheshmehzangi, *Urban Health, Sustainability, and Peace in the Day the World Stopped*, Sustainable Development Goals Series, https://doi.org/10.1007/978-981-16-4888-5_10

absence or malfunctioning in the time of need, and unhealthy relationships that may exist between institutions and the general public.

While a significant part of corruption is linked with social and economic development (Morse 2006), some scholars also suggest the impacts it may have on environmental sustainability (Lopez and Mitra 2000, Damania et al. 2003, Morse 2004; Welsche 2004; Morse 2006). Thus, economic growth and economic stability are tangled with lower corruption levels, assuring societal health and economic sustainability are facilitated through socio-economic goods of the context. At all levels, we see the relationship between a low level of sustainability and a high level of corruption (Khlif et al. 2016), suggesting the complexity that exists and affects countries, regions, cities, and communities. Some of the impacts are related to the choice of index (Ewers and Smith 2007). Still, some are primarily associated with relationships that exist between institutions, monitory systems, and mechanisms that reduce common corruption practices.

In this chapter, we look at one of the main positive peace pillars, i.e. a low level of corruption. Instead of exploring it from the institutional or political perspective, we explore this dimension from the economic sustainability perspective. In doing so, we aim to consider the health–sustainability–peace (HSP) nexus from the economic stability and sustainability perspectives, recognising the importance of low-level corruption in maintaining the health and healthiness of economies. The economic sustainability of the economy itself is considered as a major indicator in enhancing distributional justice and combating corruption (Spangenberg 2005), which in a way help to strengthen "*the respect for societal conflict management mechanisms*" (ibid) and towards sustainable economic growth (Kim et al. 2017). Thus, we ought to realise the central position low level of corruption plays in maintaining a comprehensive HSP nexus in cities and communities.

10.2 Reflecting on How Cities Reacted During the Pandemic

The COVID-19 impacts on the economy of cities and communities have been widespread, disabling many functionalities and reducing their operational qualities (Cheshmehzangi 2020a, b, c). Depending on their contexts, some communities have experienced transitions, and some have gone through transformation processes (Cheshmehzangi 2021a). The severe recession caused by the pandemic (Shan et al. 2021) suggests larger-scale complexities such as those affecting the socio-economic needs of societies around the globe (Andreoni 2021). In addition, we see significant regional differences (Cheshmehzangi et al. 2021) that led to earlier inactions or incapability in managing and containing the outbreak. Some of these are studied from the correlated economic, geographical, and facilities perspectives, suggesting how cities/regions could have reacted more responsibly if economic stability was in place.

Moreover, the COVID-19 pandemic has highlighted the increasing risk of corruption activities during such events (Terziev and Georgiev 2020). Some studies highlight the effect of public corruption on increasing infection and death rates during the pandemic (Farzanegan 2021), while others relate the issues with political connections (Gallego et al. 2020). In all cases, we see growing institutionalised corruption that led to the immediate decline of an ineffective response to the pandemic (Mietzner 2020). By questioning the role of country-level prosperity in managing the COVID-19 pandemic, Khorram-Manesh et al. (2020) suggest that the risk of the pandemic appeared much higher in high prosperity communities, which could indicate that a high-level economy alone cannot really be the mere answer to better stability conception. In fact, we could note potential conflicts in such contexts that could lead to widening the societal gap and higher-level vulnerabilities in different social classes of society.

The other aspects worth mentioning are issues of personalised power in the name of minimising economic disruptions (Ang 2020) and neglecting the right attitude towards effective decision-making. The follow-up economic deprivation caused by such attitudes showed a larger-scale impact on economies that are already poor (Khan and Roy 2020), also widening the gap between citizens and their immediate governments (Fisman and Golden 2017). While many civil law agreements are in place to reduce or prevent corruption within the pandemic (Hamzah 2020), we see a growing number of cases that show corruption in public institutions (Khemani 2020) and managing the economies. More importantly, we see how choices were made regarding achieving stability of economic–institutional sustainability during the early stages of the pandemic. From early response and management plans to the provision of support to public funds, social benefits, taxation support, etc., we see growing issues that widen the gaps between the low-income groups and people at or close to the top of the pyramid. As the COVID-19 pandemic overwhelms low-income countries (Ali and Ali 2020), we see agonising—yet expected—consequences of the COVID-19 on low-income people. The combined health and economic impacts often are the result of lower-level economic stability, affecting more the vulnerable institutions that may not perform to the expected standards. Thus, we see higher health risk exposure and higher economic decline in places where economies were already corrupted or dysfunctional.

However, evidence shows that some cities and communities truly managed to react responsibly and reflectively in containing the COVID-19 pandemic (Cheshmehzangi 2020d). The upright use of infrastructure (D'Adamo and Rosa 2020) is an example of how cities reflected to enhance their public–private partnerships and enhance the sectoral and cross-sectoral opportunities essential for making the operations more resilient and effective (Cheshmehzangi 2020d). We also note the effects through transparent and unpoliticised collaborative approaches (Saleh et al. 2021), suggesting a healthier administration capacity, better economic support, and a more sustainable response plan during the pandemic.

While uncertainties led to larger impacts on welfare economies (Rose-Ackerman 2021), we saw how important it was to have good institutional and economic development (Aidt 2009) prior to preparing the cities and communities against the pandemic. The economic crisis that was experienced at first was mostly related to structural causes (Ozili 2020) that goes beyond just the global economic issues. Some of the impacts resulted from political distrust (Ezeibe et al. 2020), knowing that economies are not stable due to political problems and socio-economic conditions, such as lack of management mechanisms, and monitory and control frameworks. In many cases, the corruption level relates to the context and its economic drivers (Dincer and Gillanders 2021); hence, we also explore it from the perspective of economic sustainability. Our political and economic systems are correlated in many ways that could not be separated so easily. In the so-called democratic states, we see the hidden ways that try to separate the two. At the same time, it is evident that economic and political institutions operate in the same way and are usually the same thing. Hence, regardless of the attempts, we see much more significant issues that are sold in the name of democracy and freedom, while in fact, are operational in the benefit of political systems. We see the same that has happened during the COVID-19 pandemic while affecting the stock market, gas prices, GDP changes, industrial production, etc. Thus, it is only naïve to think about the difference between economic and political institutions. The same also applies to cities, where the city authorities are direct beneficiaries of economic operations. The city's GDP and other economic indicators often are correlated with their political status and conditions, meaning that the two are inseparable.

Atiles Osoria (2021) studies the case of Puerto Rico, which is a poor state of the USA in colonial settings. It is evidenced that places like Puerto Rico continuously lack economic support from their upper governments, meaning a higher risk of exposure to health crises and emergencies. His study also investigates state-corporate crimes that cause more adversities in maintaining economic health and sustainability. The same applies to countries that have poorer provinces, regions, and communities. In such cases, the extent of poverty cannot be limited based on the country's GDP or another economic index, but the disparities and structural changes that exist in society (Cheshmehzangi 2021b). The socio-economic inequalities are genuinely the ones that led to longer corruption situations and creating the opportunity for multiple standards, processes, and responses. Unfortunately, in most cases, with some exceptions, the attention is not given to those at the bottom. Even if they are the essential workers, they may still feel more vulnerable because of unbalanced economic support.

10.3 Addressing the Flaws

Saleh et al. (2021) mentioned that transparency and unpoliticised collaborative approaches are crucial to reducing corruption and reversing dysfunctional systems' operations. The severe economic crisis caused by the COVID-19 pandemic shows that some prosperous countries faced significant issues not because of the lack of economic prosperity but because of a higher level of hidden corruption, large-scale socio-economic inequalities, and widening structured dysfunctionalities in essential institutions. In such prosperous contexts, the success is on individual or marginal prosperity, but the so-called public prosperity is far from reach. To address these more responsibly, we may need to revisit the structural and institutional systems that determine the backbone of our economic stability. With an open eye, we have to realise the issues that restrict our integrative multidisciplinary and multisectoral approaches and those collaborative opportunities that simply make sense. We have to consider the unhealthy nature of growing competitiveness in the face of collaboration, combatting, in particular, the issues that may lead to a higher level of corruption or concealed relationships and diplomacies behind the scenes. We also need to have a better understanding of economic sustainability and stability as better means to never-ending economic growth targets.

In order to suggest constructive thoughts to combat corruption and not ending up behind jail bars, we need to be reflective and honest about the health and stability of society. We ought to see a better picture of socio-economic inequalities that are born from larger-scale global economic corruptions and suggest ways out of longer-term structured corruptions. If the growing global widening gap does not enlighten us of the situations that have—for long—become the (common) norm. We may simply neglect the corrupted economies that increase the inequalities within and in between societies. The disproportionate poverty often exists in cities or between cities of the same region because competitiveness exists and creates unhealthy conditions of limited collaboration, poor quality cooperation, and injustice societies. To achieve low-level corruption in economic development and stability, we ought to revisit our institutional structures and invest more in public–private partnerships (PPP), small- and medium-sized enterprises (SMEs), and people-oriented economic sustainability. To address these, we may need to consider a balance between public health and economic prosperity (Hodge et al. 2020) and protect prosperity (Pak et al. 2020) against unsustainable patterns of economic growth. As mentioned in the previous chapter, we also need to consider business sustainability from a system perspective (Bansal et al. 2021) and ensure social prosperity is a long-term plan (Bartik 2020; Cheshmehzangi 2020d), and not a short-term idea. The situation, as it persists, also indicates a lack of commercial prosperity (Li et al. 2021), showing the horizontal turn in governance (Horton 2020) than just vertical growth. The economic health and stability that are most welcomed in cities and communities ought to help strengthen a wider range of urban systems and protect the networks of health, prosperity, and security (Cheshmehzangi 2020d; Loembé et al. 2020). Thus, we ought to ensure financial stability, better preparedness, and monitory in developing less corrupted economies, better social stabilities, and healthier institutional structures.

10.4 A Summary

In this chapter, we highlighted an important part of the HSP nexus, i.e. a low level of corruption in economic sustainability. As thoroughly discussed, the "political will" behind lower-level corruption is crucial to societal prosperity. The issue of corruption does not really separate developed and developing countries so much but highlights the lack of support and institutions in developing nations/contexts. Corruption is inevitably part of human nature, and without the control and monitory mechanisms, the end results may not differ much. Thus, our progress towards lowering the corruption levels should be longitudinal and disruptive in a way that operations could be monitored, checked, and sustained. The unhealthy patterns of development suggest growing corruption opportunities not only for governments and institutions but also for larger businesses, international organisations, NGOs, charities, and others. Evidence shows that during the COVID-19 pandemic, there were cases of corruption at all levels and in all sorts of organisations. Thus, it is only reasonable for us to consider combining governance and economic sustainability to ensure socio-economic prosperity is a long-term plan. To ensure peace, we have to recognise the value of economic sustainability and not just economic growth. It is critical to know that the growth on its own does not resemble sustainability, particularly that it has become a tool to widen the gaps rather than making collaborations or integrative solutions.

To summarise briefly, we reflect on the institutional–economic correlation in reducing corruption levels and towards economic sustainability. Urban systems are playful parts of urban economies and their stability. Thus, it is vital for them to be a collaborative part of a healthy urban ecosystem, a symbiosis across sectors and between different social groups, regions, and environments. To further expand on the symbiosis conception, the next chapter delves into the role of the supply chain as a suitable example under economic sustainability and achieving the HSP nexus in cities and communities.

References

Aidt TS (2009) Corruption, institutions, and economic development. Oxf Rev Econ Policy 25(2):271–291

Ali I, Ali S (2020) Why may COVID-19 overwhelm low-income countries like Pakistan?. Disaster Med Public Health Preparedness 1–5

Andreoni V (2021) Estimating the European CO_2 emissions change due to COVID-19 restrictions. Sci Total Environ 769:145115

Ang YY (2020) When COVID-19 meets centralized, personalized power. Nat Hum Behav 4(5):445–447

Atiles Osoria, J. (2021). The COVID-19 pandemic in Puerto Rico: exceptionality, corruption and state-corporate crimes. State Crime J 10(1):104–125

Bansal P, Grewatsch S, Sharma G (2021) How COVID-19 informs business sustainability research: It's time for a systems perspective. J Manage Stud 58(2):602–606

Bartik TJ (2020) Helping America's distressed communities recover from the COVID-19 recession and achieve long-term prosperity. Policy Brief. Washington, DC: Brookings Metropolitan Policy Program. https://www.brookings.edu/research/helping-americas-distressedcommunities-recover-from-the-covid-19-recession-and-achieve-long-term-prosperity

Cheshmehzangi A (2020a) Recommendations for 'The city in need.' In: Cheshmehzangi A (ed) The city in need: urban resilience and city management during disruptive disease outbreak events. Springer, Singapore, pp 285–304

Cheshmehzangi A (2020b) Reflection on disruptions: managing the city in need, saving the city in need. In: Cheshmehzangi A (ed) The city in need: urban resilience and city management during disruptive disease outbreak events. Springer, Singapore, pp 137–283

Cheshmehzangi A (2020c) Reflection on early lessons for urban resilience and public health enhancement during the COVID-19. Health 12(10):1390

Cheshmehzangi A (2020d) The city in need: urban resilience and city management during disruptive disease outbreak events. Springer, Singapore

Cheshmehzangi A (2021a) From transitions to transformation: a brief review of the potential impacts of COVID-19 on boosting digitization, digitalization, and systems thinking in the built environment. J Build Constr Plann Res 9(01):26

Cheshmehzangi A (2021b) Vulnerability of the UK's BAME communities during COVID-19: the review of

public health and socio-economic inequalities. J Human Behav Soc Environ 1–18
Cheshmehzangi A, Sedrez M, Ren J, Kong D, Shen Y, Bao S, Xu J, Su Z, Dawodu A (2021) The effect of mobility on the spread of COVID-19 in light of regional differences in the European Union. Sustainability 13(10):1169794
Chinweuba GE (2020) Low level corruption in Nigerian society: a critical investigation. OWIJOPPA 4(1):26–40
D'Adamo I, Rosa P (2020) How do you see infrastructure? Green energy to provide economic growth after COVID-19
Damania R, Fredriksson PG, List JA (2003) Trade liberalization, corruption, and environmental policy formulation: theory and evidence. J Environ Econ Manag 46(3):490–512
Ewers RM, Smith RJ (2007) Choice of index determines the relationship between corruption and environmental sustainability. Ecol Soc 12(1)
Ezeibe CC, Ilo C, Ezeibe EN, Oguonu CN, Nwankwo NA, Ajaero CK, Osadebe N (2020) Political distrust and the spread of COVID-19 in Nigeria. Glob Public Health 15(12):1753–1766
Farzanegan MR (2021) The effect of public corruption on COVID-19 fatality rate: a cross-country examination. CESifo Working Paper No. 8938. Available from: https://papers.ssrn.com/sol3/papers.cfm?abstract_id=3805464
Fisman R, Golden M (2017) How to fight corruption. Science 356(6340):803–804
Gallego JA, Prem M, Vargas JF (2020) Corruption in the times of pandemia. Available at SSRN 3600572
Haass F, Ottmann M (2017) Profits from peace: the political economy of power-sharing and corruption. World Dev 99:60–74
Hamzah H (2020) Civil law agreement and its implication on regulation for prevention of corruption within Covid-19 Pandemic. J Soc Stud Educ Res 11(3):156–176
Hodge Jr, JG, Wetter S, Carey E, Pendergrass E, Reeves CM, Reinke H (2020) Legal "tug-of-wars" during the COVID-19 pandemic: public health versus economic prosperity. J Law, Med Ethics 48(3):603–607
Horton R (2020) Offline: COVID-19—a crisis of power. The Lancet 396(10260):1383
Jain AK (ed) (2001) The political economy of corruption, vol 2. Routledge, New York and London
Khan M, Roy P (2020) COVID-19: locking in solutions while in lockdown. Available from: https://eprints.soas.ac.uk/33223/1/ACE-BriefingPaper008-Covid19-Locking-in-solutions-while-in-lockdown-200505.pdf
Khemani S (2020) An opportunity to build legitimacy and trust in public institutions in the time of COVID-19. World bank research and policy briefs, (148256). Available from: https://papers.ssrn.com/sol3/papers.cfm?abstract_id=3602352#maincontent
Khlif H, Guidara A, Hussainey K (2016) Sustainability level, corruption and tax evasion: a cross-country analysis. J Fin Crime 23(2). Available from: https://doi.org/10.1108/JFC-09-2014-0041/full/html?casa_token=Wl2yiPl-EAsAAAAA:8CkfSJhZvoMYCGH6FkvLkf4-b2WLt_Ytm6KibUvLbDURNUykfrUSCcfSoLfgMvaJ7N4z69sAyOy2-U8OceyKpIx4Q36KtMpmmrD7VHx25WFIseRVhYY
Khorram-Manesh A, Carlström E, Hertelendy AJ, Goniewicz K, Casady CB, Burkle FM (2020) Does the prosperity of a country play a role in COVID-19 outcomes?. Disaster Med Public Health Preparedness 1–10
Kim E, Ha Y, Kim S (2017) Public debt, corruption and sustainable economic growth. Sustainability 9(3):433
Le Billon P (2003) Buying peace or fuelling war: the role of corruption in armed conflicts. J Int Dev: J Dev Stud Assoc 15(4):413–426
Le Billon P (2008) Corrupting peace? Peacebuilding and post-conflict corruption. Int Peacekeeping 15(3):344–361
Li B, Peng Y, He H, Wang M, Feng T (2021) Built environment and early infection of COVID-19 in urban districts: a case study of Huangzhou. Sustain Cities Soc 66:102685
Loembé MM, Tshangela A, Salyer SJ, Varma JK, Ouma AEO, Nkengasong JN (2020) COVID-19 in Africa: the spread and response. Nat Med 26(7):999–1003
Lopez R, Mitra S (2000) Corruption, pollution, and the Kuznets environment curve. J Environ Econ Manag 40(2):137–150
McMann KM, Seim B, Teorell J, Linberg SI (2019) Why low levels of democracy promote corruption and high levels diminish it. Polit Res Q 73(4). https://doi.org/10.1177/1065912919862054
Mietzner M (2020) Populist anti-scientism, religious polarisation, and institutionalised corruption: How Indonesia's democratic decline shaped its COVID-19 response. J Curr Southeast Asian Aff 39(2):227–249
Morse S (2004) Indices and indicators in development. An unhealthy obsession with numbers? Earthscan, London, UK
Morse S (2006) Is corruption bad for environmental sustainability? A cross-national analysis. Ecol Soc 11(1):22. Available from: http://www.ecologyandsociety.org/vol11/iss1/art22/
Newham G (2014) Why is corruption getting worse in South Africa. Corruption watch. January, 17. Available from: www.hsrc.ac.za/en/hsrc-in-the-news/dces/corruption-0114
Ozili PK (2020) Covid-19 pandemic and economic crisis: The Nigerian experience and structural causes. J Econ Adm Sci. Available from: https://doi.org/10.1108/JEAS-05-2020-0074/full/html?casa_token=3xafef4_IPMAAAAA:Dvy_DW7lqjYzWebF9Ae8x2X8V-D6uJM2xidRSXKxSxCn9lcLwCpnnst0Ji9Q9D1aebK4DOewbPMTSmyKTJpFWuAQup2fqVIVWgbzRC8DulN5zM4BJvYY

Pak A, Adegboye OA, Adekunle AI, Rahman KM, McBryde ES, Eisen DP (2020) Economic consequences of the COVID-19 outbreak: the need for epidemic preparedness. Front Public Health 8

Rose-Ackerman S (2008) Corruption and post-conflict peace-building. Ohio NUL Rev 34:405

Rose-Ackerman S (2021) Corruption and COVID-19. EUNOMÍA Revista En Cultura De La Legalidad 20:16–36

Saleh S, Naal H, Mokdad AH (2021) It's all in the details: a call for administering the COVID-19 vaccine in Lebanon through a transparent and un-politicized collaborative approach. EClinicalMedicine 32

Shan Y, Ou J, Wang D, Zeng Z, Zhang S, Guan D, Hubacek K (2021) Impacts of COVID-19 and fiscal stimuli on global emissions and the Paris agreement. Nat Clim Chang 11(3):200–206

Spangenberg JH (2005) Economic sustainability of the economy: concepts and indicators. Int J Sustain Dev 8 (1–2):47–64

Terziev V, Georgiev M (2020) Increasing the risk of corruption activities during a COVID-19 pandemic. Available at SSRN 3674427

Welsch H (2004) Corruption, growth, and the environment: a cross-country analysis. Environ Dev Econ 9:663–693

11 The Role of Supply Chain Resilience in Contemporary Age

The reality is often what we cannot see. So, be well prepared.
—Ali Cheshmehzangi, 2016

11.1 A Brief Exploration: Supply Chain Resilience and Healthy Economies

In this chapter, we explore the notion of the supply chain in the contemporary age, which we believe exists for a long time in cities and communities throughout history. In their study on two perspectives on supply chain resilience, Wieland and Durach (2021) highlight monitory global supply chains and management factors regarding strategic management and assessment opportunities for productions. It is also noted that disruption and competition do coexist side-by-side in a dynamic urban supply chain (Friesz et al. 2011), allowing urban consolidation centres to operate as a response to efficient operations and generations (Allen et al. 2014). In some cases, hybrid models have been introduced, ensuring healthy hybrid supply chain strategies to face the challenges of disruptive times (Nakandala and Lau 2019). The end result ensures a sustainable urban supply chain is achieved (Graham et al. 2015) in a process that is fair, all-inclusive, and well-functioning (Cheshmehzangi 2020a). Thus, we often consider a risk management approach to supply chain sustainability (Giannakis and Papadopoulos 2016), with specific metrics (Clift 2004), and specific management practices (Govindan et al. 2014; Martins and Pato 2019). The considerations have been utilised comprehensively across sustainability dimensions (Mota et al. 2015) and in cooperation with innovation framework opportunities (Kusi-Sarpong et al. 2019), supply chain orientation (Jadhav et al. 2019), and benchmarking opportunities (Colicchia et al. 2011). Some of these areas have developed new streams of research, correlating supply chain sustainability with resilience (Fahminia and Jabbarzadeh 2016) and towards risk assessment and management scenarios (Xu et al. 2019).

Supply chain resilience itself is a concept that helps to reduce risks and minimise issues that may cause disruptions in supply chain networks and/or clusters in cities (Ponis and Koronis 2012). Its evolution in supply chain management has helped highlight new resilience factors, which also respond well to potential political conflicts, economic instability, and disruptions (Pettit et al. 2019). A good example is urban food security issues (Burton et al. 2013) or related sustainability factors that may highlight more cautiously the effects of supply chain resilience as a whole (Hecht et al. 2019; Langemeyer et al. 2021). In dealing with the unpredictable, supply chain resilience requires adequate conceptuali-

A. Cheshmehzangi, *Urban Health, Sustainability, and Peace in the Day the World Stopped*, Sustainable Development Goals Series, https://doi.org/10.1007/978-981-16-4888-5_11

sation and scale development (Scholten et al. 2019), guiding resilience policy design (Hecht et al. 2019). For instance, in mitigation processes (Scholten et al. 2014), we see a higher opportunity for economic restructuring, assessment, and post-disaster analysis (Dixit et al. 2020) and towards better performances in supply chain networks (Li et al. 2017). The existing scholarly work also refers to supply chain resilience as an assessment tool and an implementation process (Pettit 2008). Thus, we see excellent opportunities to manage risks and vulnerability through the proper decision-making processes (Wu and Blackhurst 2009) and economic innovation. In particular, we note the network of operations and production as part of the more extensive network of economic sustainability (Ivanov 2018), suggesting more than just an assessment for disruptive times, but an evolution for better urban management opportunities.

While some argue that a better understanding of resiliency in supply chains is not possible (Ponomarov and Holcomb 2009; Novak et al. 2021), we note the strong nexus between the supply chain and sustainability. In particular, this applies to dynamic contexts where the shift is towards a circular economy (Bag et al. 2019), sustainable development, and other types of reducing the negative impacts of recent and contemporary development. To ensure supply chain resilience is successful, one ought to define the metrics, target population, and demand volatility of supply chain networks (Pettit et al. 2013). In doing so, we see better chances for understanding the daily operations and whether they are healthy or resilient at all. It is, therefore, crucial to evaluate how supply chain resilience could feed into the economic sustainability of cities and communities. In this regard, this chapter looks into supply chain issues and particularly during the pandemic events like the ongoing COVID-19 pandemic. The explanation on noticeable disruptions of the pandemic on supply chain networks is not repeated here as we aim to highlight more important factors, such as their resilience, collaboration, and sustainability during the course of pandemic crises.

11.2 Reflecting on How Cities Reacted During the Pandemic

Early lessons from the COVID-19 pandemic suggest the issues of urban resilience and public health (Cheshmehzangi 2020b). While much scholarly work identifies a broad range of economic impacts of the pandemic (Abiad et al. 2020; Atkeson 2020; Baker et al. 2020; Che et al. 2020; Cheshmehzangi 2020c; Hevia and Neumeyer 2020; Maital and Barzani 2020; McKibbin and Fernando 2020; Park et al. 2020), little knowledge is added in terms of achieving economic sustainability through specific mechanisms, such as supply chain systems. The so far research work highlights the viability of intertwined supply networks (Ivanov and Dolgui 2020), extending resilience research towards survivability and developing integrity, security, and provision of services. There are, however, substantial gaps between research findings and industry practice, which are highlighted by Remko (2020) through the study of the post-COVID-19 supply chain. Specific sectors, such as manufacturing and services operations, suggest examples of economic sustainability mechanism (Belhadi et al. 2021), knowing a wide range of emerged and emerging issues in relation to global supply chains (Ivanov and Das 2020) and confronting disruptions caused by disasters and emergencies like the pandemic.

The supply chain disruptions on cities and urban systems are widespread and create a variety of adversities on city operations (Agrawal et al. 2020; Mahajan and Tomar 2020; Cheshmehzangi et al. 2021a; Mahajan and Tomar 2021). Some examples are the impact on logistic systems that caused food supply chain disruptions (Singh et al. 2021), showing the inefficiency in local and regional production. Most cities experienced at least a minimum of two to four weeks of disruption, which may be considered normal. Nevertheless, anything beyond just a few weeks shows fragility in managing the supply change, its distribution, and resilience. The consideration of adaptive planning in supply chain management was an early sign of enhancing the operations, productions, and production generations (Cheshmehzangi 2020a). Thus, we see the importance of supply chains and economic impacts (Inoue and Todo 2020), at least from the actual supply change linkages and connectivity between cities, communities, and even regions.

As cities shifted towards adaptive measures (Cheshmehzangi 2020a), we note opportunities that were also raised in specific areas, such as additive manufacturing (Larrañeta et al. 2020), adaptive leadership (Garavaglia et al. 2021), and towards adaptive management of spaces and infrastructures (Cheshmehzangi 2020d; Suriadi 2021). Adaptive control measures indicate only temporary solutions to longer-term economic issues, while some research also indicates issues related to movement, mobility, connectivity, etc. (Malani et al. 2020; Sharifi and Khavarian-Garmsir 2020). We also note that supply chain resilience was tested during the early stages of the pandemic, leading to a later adaptive approach that tried to stabilise the larger-scale impacts. Nonetheless, the closure of many retail and commercial units and supplies showed the vulnerability of smaller businesses against giant enterprises and companies. The longer-term effects show inequalities that exist in economic productions and within the business environments. The unbalanced situation between larger profit-making companies and SMEs, for instance, indicates that supply chain resilience is not the same between different sectors and businesses of different scales.

11.3 Addressing the Flaws

If cities rely mainly on external supply chain networks, as they mostly are, then they become more vulnerable to situations like the COVID-19 pandemic or other crisis events. The lack of regional planning in such cases could heavily affect the imbalanced production and consumption trends, such as energy (gas, electrical, etc.), water, and food. We also see similar issues growing in other sectors, such as manufacturing, where we note more significant impacts on supply chain agility (Al-Zabidi et al. 2021) and chain reactions (Krellenberg and Koch 2021). As demonstrated by Hong and Choi (2021), the example of high economic resilient neighbourhoods shows us many benefits of specific urban characteristics that could provide better socio-economic determinants for healthier communities. To start with, we have to deal with long-term inequalities (Cheshmehzangi 2021b) that have become the main determinants in defining urban communities across the globe. We ought to see the importance of supply chain networks and their resilience in maintaining optimal financing strategies and cautiously towards a capital-constrained supply chain (Zhang et al. 2021). Examples of these could be seen in assessing urban energy resilience (Sharifi and Yamagata 2016), food security measures, and other metrics of supply chain resilience (Han et al. 2020). We could then note the innovative opportunities that exist in understanding adaptive approaches, creating healthy mechanisms for better supply chain networks and resilience.

In addition, during the COVID-19 pandemic, the impacts on specific sectors have been perceived more widely by the general public. Such consequences show the unsustainability of supply chains in the name of economic growth and prosperity, pressuring the most vulnerable groups in particular. The lack of self-sufficiency in supply chain mechanisms also indicates longer-term issues that create global supply chain insecurity and add to other thresholds such as high economic targets and lack of self-sufficient productions. Thus, building resilience in supply chain management has been highlighted as a significant feasible approach to overcome some issues in supply chain functionality, value chain (s), and partnerships (Falsetta 2020). More innovative ideas, such as utilising the Internet of Things (IoT) to design a relief supply chain network (Dutta and Mitra 2021; Zahedi et al. 2021), are potential methods that we could see developing more out of the current pandemic.

To summarise some of the critical suggestions, we note addressing the flaws from the perspective of self-sufficiency. This thinking, however, may not be feasible for every context. Thus, the poorer economies should consider approaches of regional development that may need to go beyond their (own) political boundaries and towards sub-regional mechanisms for cooperation, integrative collaboration, and joint economic sustainability goals. While we still believe in the benefits of competitiveness, we also note the issues it may have on managing economies and dependencies. The multilateral opportunities could create a better and healthier regional development, which could also address other sustainability factors that just those of economic nature. For economic growth to become steady and sustainable, we ought to understand the values and mechanisms of collaboration especially that we see the adverse impacts of emergencies or health crises could be beyond just closure of boundaries.

During the early phases of the COVID-19 pandemic, as more prosperous nations/regions opt for their by-default/business-as-usual practices, the impact was seen in their own lands and their neighbouring states. This issue was seen in specific regions regardless of their economic status, from North America to South America, South East Asia to the Middle East, and so on. To ensure peace is maintained, we note the importance of humanitarian logistics to better manage supply chain disruptions, which is remarkably discussed by Kovacs and Falagara Sigala (2021). Of course, some conflicting points are highlighted by Rossi and Mingardi (2020), showing the broader issues of cybersecurity, personal data, sensitive information, supply chain reports, etc. Thus, the city governments and authorities should ensure they provide the goodwill to respond to the needs of the city, avoiding the abuse of a peaceful environment in the peacebuilding process.

Lastly, we highlight the ever-changing dynamism that exists in our economic growth, development, and stability. We note the correlations between local economies and globalised economic development patterns, particularly from the supply chain perspectives. Scholars like Barry (2004), who evaluate the scope of supply chain sources, patterns of development, and markets, have discussed the issues of "*supply chain risk in an uncertain global supply chain environment*". Other scholars also touch on key strategic opportunities, especially during the COVID-19 pandemic. These include other areas that would fit well with supply chain systems, such as economic development mechanisms, global health measures (Chattu and Chami 2020), and better supply chain management. Therefore, we foresee some adjustments and revisions to what we currently have in our global trade and supply chain systems, ensuring that countries, regions, and cities collaborate with each other but are also self-sufficient, vibrant, and sustainable.

11.4 A Summary

As this chapter closes the economic dimension of the book, we suggest a wide range of reflections on economic factors related to the health–sustainability–peace (HSP) nexus of cities and communities. More importantly, as shown in this chapter, we ought to recognise the value of adaptable systems (Amadei 2021), more likely for understanding not only the research agenda for sustainable development (Ranjbari et al. 2021) but also practical pathways towards achieving the sustainability goals. The ongoing COVID-19 pandemic may help us redefine the concept of sustainability (Hakovirta and Denuwara 2020), at least hoping that sustainability discussion could differ in the future, and factors such as economic growth could mean differently and towards economic stability and sustainability. In this chapter, we delve into some of the key arguments around supply chain networks and management, some that truly reflect on issues of resilience and sustainability of such complex systems. The aim here is to ensure we are not just focused on economic recovery but to revisit some of the economic issues that have led to unsustainable trends and practices and towards transformative responses. Nonetheless, the impacts are mainly on society and people, especially those of the bottom of the pyramid, the (often forgotten) vulnerable groups, and the marginalised and poor populations. Thus, the following three chapters address the sister dimension to the economic dimension, i.e. the social dimension of the HSP nexus in cities and communities.

References

Abiad A, Arao RM, Dagli S (2020) The economic impact of the COVID-19 outbreak on developing Asia. ADB briefs, Asian Development Bank, http://dx.doi.org/10.22617/BRF200096, Available from: https://www.adb.org/publications/economic-impact-covid19-developing-asia

Agrawal S, Jamwal A, Gupta S (2020) Effect of COVID-19 on the Indian Economy and supply chain. Available from: https://www.preprints.org/manuscript/202005.0148/v1

Allen J, Browne M, Woodburn A, Leonardi J (2014) A review of urban consolidation centres in the supply chain based on a case study approach. In: *Supply chain forum: an international journal*, vol. 15, No. 4. Taylor & Francis, pp 100–112

Al-Zabidi A, Rehman AU, Alkahtani M (2021) An approach to assess sustainable supply chain agility for a manufacturing organization. Sustainability 13 (4):1752

Amadei B (2021) A systems approach to the sustainability–peace nexus. Sustain Sci 1–14

Atkeson A (2020) What will be the economic impact of COVID-19 in the US? Rough estimates of disease scenarios (No. w26867). National Bureau of Economic Research

Bag S, Gupta S, Foropon C (2019) Examining the role of dynamic remanufacturing capability on supply chain resilience in circular economy. Manage Decis

Baker SR, Bloom N, Davis SJ, Terry SJ (2020) Covid-induced economic uncertainty (No. w26983). National Bureau of Economic Research

Barry J (2004) Supply chain risk in an uncertain global supply chain environment. Int J Phys Distrib Logist Manag

Belhadi A, Kamble S, Jabbour CJC, Gunasekaran A, Ndubisi NO, Venkatesh M (2021) Manufacturing and service supply chain resilience to the COVID-19 outbreak: Lessons learned from the automobile and airline industries. Technol Forecast Soc Change 163:120447

Burton P, Lyons K, Richards C, Amati M, Rose N, Desfours L, Pires V, Barclay R (2013) Urban food security, urban resilience and climate change. National Climate Change Adaptation Research Facility, Australia. Available from: https://eprints.qut.edu.au/68195/

Chattu VK, Chami G (2020) Global health diplomacy amid the COVID-19 pandemic: a strategic opportunity for improving health, peace, and well-being in the CARICOM Region—a systematic review. Soc Sci 9 (5):88

Cheshmehzangi A (2020a) The City in need: urban resilience and city management during disruptive disease outbreak events. Springer, Singapore

Cheshmehzangi A (2020b) Reflection on early lessons for urban resilience and public health enhancement during the COVID-19. Health 12(10):1390

Cheshmehzangi A (2020c) COVID-19 and household energy implications: what are the main impacts on energy use?. Heliyon 6(10):e05202

Cheshmehzangi A (2020d) 10 Adaptive measures for public places to face the COVID 19 pandemic outbreak. City Soc 32(2)

Cheshmehzangi A (2021a) From transitions to transformation: a brief review of the potential impacts of COVID-19 on boosting digitization, digitalization, and systems thinking in the built environment. J Building Constr Planning Res 9(01):26

Cheshmehzangi A (2021b) Vulnerability of the UK's BAME communities during COVID-19: the review of public health and socio-economic inequalities. J Hum Behav Soc Environ 1–18

Clift R (2004) Metrics for supply chain sustainability. Technol Choices Sustain 239–253

Colicchia C, Melacini M, Perotti S (2011) Benchmarking supply chain sustainability: insights from a field study. Benchmarking: Int J

Dixit V, Verma P, Tiwari MK (2020) Assessment of pre and post-disaster supply chain resilience based on network structural parameters with CVaR as a risk measure. Int J Prod Econ 227:107655

Dutta PK, Mitra S (2021) Application of agricultural drones and IoT to understand food supply chain during post COVID-19. Agric Inf Autom Using IoT Mach Learn 67–87

Fahimnia B, Jabbarzadeh A (2016) Marrying supply chain sustainability and resilience: a match made in heaven. Trans Res Part E: Logistics Trans Rev 91:306–324

Falsetta M (2020) Building resilience in supply chain management: the post Covid-19 experience in BtoB manufacturing firms. Available from: tesi.cab.unipd.it/65476/1/Falsetta_Matteo.pdf

Friesz TL, Lee I, Lin CC (2011) Competition and disruption in a dynamic urban supply chain. Trans Res Part B: Methodological 45(8):1212–1231

Garavaglia C, Sancino A, Trivellato B (2021) Italian mayors and the management of COVID–19: adaptiveleadership for organizing local governance. Eurasian Geogr Econ 62(1):76–92

Giannakis M, Papadopoulos T (2016) Supply chain sustainability: a risk management approach. Int J Prod Econ 171:455–470

Govindan K, Azevedo SG, Carvalho H, Cruz-Machado V (2014) Impact of supply chain management practices on sustainability. J Clean Prod 85:212–225

Graham G, Österle I, Aditjandra PT, Vaghi C, Grea G, Zunder TH (2015) The role of a structured stakeholder consultation process within the establishment of a sustainable urban supply chain. Supply Chain Manage An Int J 20(3). Available from: https://www.emerald.com/insight/content/doi/10. 1108/SCM-05-2014-

0149/full/html?casa_token=K0lM6RzSvNkAAAAA:Ceymh6nCidXG0uxdhZO4TZ3XOyCtOvyRAUvTeVjiEkj_ysufeOwXUTIJMCWEh-_IinCozo0eN_exxQ-kCs2RP70NH3PK_Rlh9mE4NCyHDRqKMLnFzRo
Hakovirta M, Denuwara N (2020) How COVID-19 redefines the concept of sustainability. Sustain 12 (9):3727. https://doi.org/10.3390/su12093727
Han Y, Chong WK, Li D (2020) A systematic literature review of the capabilities and performance metrics of supply chain resilience. Int J Prod Res 58(15):4541–4566
Hecht AA, Biehl E, Barnett DJ, Neff RA (2019) Urban food supply chain resilience for crises threatening food security: a qualitative study. J Acad Nutr Diet 119 (2):211–224
Hevia C, Neumeyer A (2020) A conceptual framework for analyzing the economic impact of COVID-19 and its policy implications. UNDP LAC COVID-19 Policy Documents Ser 1:29
Hong S, Choi SH (2021) The urban characteristics of high economic resilient neighborhoods during the COVID-19 pandemic: a case of Suwon South Korea. Sustainability 13(9):4679
Inoue H, Todo Y (2020) The propagation of economic impacts through supply chains: The case of a mega-city lockdown to prevent the spread of COVID-19. PloS One 15(9):e0239251
Ivanov D (2018) Revealing interfaces of supply chain resilience and sustainability: a simulation study. Int J Prod Res 56(10):3507–3523
Ivanov D, Das A (2020) Coronavirus (COVID-19/SARS-CoV-2) and supply chain resilience: a research note. Int J Integr Suppl Manage 13(1):90–102
Ivanov D, Dolgui A (2020) Viability of intertwined supply networks: extending the supply chain resilience angles towards survivability. A position paper motivated by COVID-19 outbreak. Int J Prod Res 58 (10):2904–2915
Jadhav A, Orr S, Malik M (2019) The role of supply chain orientation in achieving supply chain sustainability. Int J Prod Econ 217:112–125
Kovacs G, Falagara Sigala I (2021) Lessons learned from humanitarian logistics to manage supply chain disruptions. J Suppl Chain Manag 57(1):41–49
Krellenberg K, Koch F (2021) Conceptualizing interactions between SDGs and urban sustainability transformations in Covid-19 times. Polit Gov 9(1):200–210
Kusi-Sarpong S, Gupta H, Sarkis J (2019) A supply chain sustainability innovation framework and evaluation methodology. Int J Prod Res 57(7):1990–2008
Langemeyer J, Madrid-Lopez C, Beltran AM, Mendez GV (2021) Urban agriculture—a necessary pathway towards urban resilience and global sustainability? Landscape Urban Plan 210:104055
Larrañeta E, Dominguez-Robles J, Lamprou DA (2020) Additive manufacturing can assist in the fight against COVID-19 and other pandemics and impact on the global supply chain. 3D Printing Addit Manuf 7(3):100–103
Li X, Wu Q, Holsapple CW, Goldsby T (2017) An empirical examination of firm financial performance along dimensions of supply chain resilience. Manag Res Rev
Mahajan K, Tomar S (2020) Here today, gone tomorrow: COVID-19 and supply chain disruptions. Forthcoming, Am J Agric Econ. Available at: https://papers.ssrn.com/sol3/Papers.cfm?abstract_id=3596720
Mahajan K, Tomar S (2021) COVID-19 and supply chain disruption: evidence from food markets in India. Am J Agr Econ 103(1):35–52
Malani A, Soman S, Asher S, Novosad P, Imbert C, Tandel V, Agarwal A, Alomar A, Sarker A, Shah D, Shen D, Gruber J, Sachdeva S, Kaiser D, Bettencourt L (2020) Adaptive control of COVID-19 outbreaks in India: Local, gradual, and trigger-based exit paths from lockdown (No. w27532). National Bureau of Economic Research
Maital S, Barzani E (2020) The global economic impact of COVID-19: a summary of research. Samuel Neaman Inst National Policy Res 2020:1–12
McKibbin W, Fernando R (2020) The economic impact of COVID-19. In: Baldwin R, Weder di Mauro B (eds) Economics in the time of COVID-19, work conducted by center for economic policy research (CEPR), CEPR Press, London, pp 45–115. Available from: https://www.incae.edu/sites/default/files/covid-19.pdf#page=52
Martins CL, Pato MV (2019) Supply chain sustainability: a tertiary literature review. J Clean Prod 225:995–1016
Mota B, Gomes MI, Carvalho A, Barbosa-Povoa AP (2015) Towards supply chain sustainability: economic, environmental and social design and planning. J Cleaner Prod 105:14–27
Nakandala D, Lau HC (2019) Innovative adoption of hybrid supply chain strategies in urban local fresh food supply chain. Suppl Chain Manag Int J 24(2). Available from: https://www.emerald.com/insight/content/doi/10. 1108/SCM-09-2017-0287/full/html?casa_token=uVxxO-rUoOYAAAAA:RCNsaEjbm-pi9opOL_q_s8QdXETTuKzn–UNWjJAyYK9tjjP-fyYbRAbagj4ebz9MPmFz1xoB3Qxr—WwcgJXwSktj016StBFSM4oCl2CNgjuWjyD06I
Novak DC, Wu Z, Dooley KJ (2021) Whose resilience matters? Addressing issues of scale in supply chain resilience. J Bus Logistics
Park CY, Villafuerte J, Abiad A (2020) An updated assessment of the economic impact of COVID-19 (No. 133). Asian Development Bank. Available from: https://www.adb.org/sites/default/files/publication/604206/adb-brief-133-updated-economic-impact-covid-19.pdf
Pettit TJ (2008) Supply chain resilience: development of a conceptual framework, an assessment tool and an implementation process. Ohio State University Columbus
Pettit TJ, Croxton KL, Fiksel J (2013) Ensuring supply chain resilience: development and implementation of an assessment tool. J Bus Logist 34(1):46–76

Pettit TJ, Croxton KL, Fiksel J (2019) The evolution of resilience in supply chain management: a retrospective on ensuring supply chain resilience. J Bus Logist 40 (1):56–65

Ponis ST, Koronis E (2012) Supply chain resilience? Definition of concept and its formative elements. J Appl Bus Res 28(5):921–935

Ponomarov SY, Holcomb MC (2009) Understanding the concept of supply chain resilience. Int J Logistics Manag

Ranjbari M, Esfandabadi ZS, Zanetti MC, Scagnelli SD, Siebers PO, Aghbashlo M, Peng W, Quatraro F, Tabatabaei M (2021) Three pillars of sustainability in the wake of COVID-19: a systematic review and future research agenda for sustainable development. J Cleaner Prod 126660

Remko VH (2020) Research opportunities for a more resilient post-COVID-19 supply chain–closing the gap between research findings and industry practice. Int J Oper Prod Manag 40(4). Available at: https://www.emerald.com/insight/content/doi/https://doi.org/10.1108/IJOPM-03-2020-0165/full/html

Rossi N, Mingardi A (2020) Italy and COVID-19: Winning the war, losing the peace? Econ Aff 40 (2):148

Scholten K, Scott PS, Fynes B (2014) Mitigation processes–antecedents for building supply chain resilience. Suppl Chain Manag Int J

Scholten K, Stevenson M, van Donk DP (2019) Dealing with the unpredictable: supply chain resilience. Int J Oper Prod Manag

Sharifi A, Khavarian-Garmsir AR (2020) The COVID-19 pandemic: Impacts on cities and major lessons for urban planning, design, and management. Sci Total Environ 142391

Sharifi A, Yamagata Y (2016) Principles and criteria for assessing urban energy resilience: a literature review. Renew Sustain Energy Rev 60:1654–1677

Singh S, Kumar R, Panchal R, Tiwari MK (2021) Impact of COVID-19 on logistics systems and disruptions in food supply chain. Int J Prod Res 59(7):1993–2008

Suriadi NA (2021) The adaptive City of Majene Infrastructure Strategy towards COVID 19 Pandemic. Bandar J Civil Eng 3(1):24–30

Wieland A, Durach CF (2021) Two perspectives on supply chain resilience. J Bus Log. Available from: https://onlinelibrary.wiley.com/doi/full/https://doi.org/10.1111/jbl.12271

Wu T, Blackhurst JV (eds) (2009) Managing supply chain risk and vulnerability: tools and methods for supply chain decision makers. Springer Science & Business Media

Xu M, Cui Y, Hu M, Xu X, Zhang Z, Liang S, Qu S (2019) Supply chain sustainability risk and assessment. J Clean Prod 225:857–867

Zahedi A, Salehi-Amiri A, Smith NR, Hajiaghaei-Keshteli M (2021) Utilizing IoT to design a relief supply chain network for the SARS-COV-2 pandemic. Appl Soft Comput 104:107210

Zhang X, Xiu G, Shahzad F, Duan Y (2021) Optimal financing strategy in a capital-constrained supply chain with retailer green marketing efforts. Sustainability 13(3):1357

Part IV

Sustainability and Peace: "Urban Health from the Social Dimension"

In the fourth part of studying sustainability and peace, we present three primary aspects representing urban health from the social dimension.

The first aspect is information sharing for a healthier society, which is also one of the main pillars of positive peace. From the social perspective, we look into this factor during the pandemic and similar health emergency crises. We note the importance of correct information sharing through the proper channels/media rather than chaotic situations that could cause social unrest and societal instability. These points are discussed in Chap. 12, mainly focused on information-sharing mechanisms. The second aspect of this part is human capital and equity, reflecting on some of the critical human resources matters. Again, this is one of the main pillars of positive peace, which is very important for social sustainability in cities and communities. The focus here would be more on the importance of human resources and equity that is needed during pandemic events. We explore these points from the social sustainability perspective. This aspect is studied in Chap. 13, where we aim to explore the role of human capital and equity in sustaining health and peace in cities. The third aspect zooms into public life's specific factor, exploring areas of the public realm and social life. As cities and communities react to situations like pandemics, their social life becomes more vulnerable and yet remains the backbone of society. This critical aspect is presented in Chap. 14, identifying various aspects of social and public life in achieving urban health, sustainability, and peace.

Information Sharing for a Healthier Society: More Than Just Big Data and Information-Based Platforms

12

Keeping people uninformed is as effective as keeping them uneducated.

—Ali Cheshmehzangi, 2019

12.1 A Brief Exploration: Information Sharing and Healthy Society

Information sharing is a suitable practice that mediates relationships between expectations and trust, allowing for enhancing trust in the long run (Butler Jr 1999). It is an area through which governing peace could be nurtured (Ilcan and Philips 2006) with an attempt to promote healthy living and a healthy society. A key aspect for consideration is an approach that could help "*deepening public–private partnerships through more robust information sharing*" (Shackelford 2011). New information could be generated and shared through such mechanisms, allowing peace opportunities to be processed throughout a given timeframe (Porat et al. 2015). Against any antitrust risks of information sharing (Anderson and Peace 2003), we note the importance of trust-building exercises and campaigns that require educating and include the general public. Once the society is part of the process through transparent information sharing, they could step up to become actors or main stakeholders of peace-building processes. All these are related directly to promoting a healthy society, ensuring power-sharing mechanisms and institutions (Hartzell and Hoddie 2003; Cammett and Malesky 2012; Gates et al. 2016; Bormann et al. 2019) that enable and promote peace-making processes.

In addition, information sharing has been practised in various sustainability research and practice. From sustainability investment (Lai et al. 2019) to sustainability benchmarking (Kerschbaum et al. 2011), we are able to assess the impact of trust and information sharing on sustainability and sustainable development (Khan et al. 2018), also allowing for better collaborative attributes, sharing services and supplies, and integrated models of sustainable development. While trust is identified to be central to information sharing (Özer et al. 2011), other concerns include issues of resource flow, cybersecurity, and operational decisions. Nonetheless, the positive aspects also highlight opportunities for better social platforms, coordination of activities, and collective decision-making processes (Kim 2018), which are indeed valuable attributes to enhance the social capital of communities and towards better social interactions (Liburd and Edwards 2010). The growing digital information sharing has become one of the recent areas of research, which has also played a major part during the COVID-19 pandemic. As suggested by Pearce (2012), the nexus between information sharing and

A. Cheshmehzangi, *Urban Health, Sustainability, and Peace in the Day the World Stopped*, Sustainable Development Goals Series, https://doi.org/10.1007/978-981-16-4888-5_12

sustainability is strong, believed to be a powerful approach to "*accelerate the transition to a sustainable world*" (ibid). Thus, we also believe in the importance of information sharing in achieving healthy social sustainability in communities and cities, especially during the calamities like health emergencies/crises, etc.

One of the key aspects in regard to information sharing in cities is related to various modes of information sharing (Ott et al. 2011), which has been studied in relation to urban mobility patterns (Vecchio and Tricarico), real-time data gathering, and use for specific areas (Guo et al. 2018), evaluating urban issues (Liu et al. 2007), etc. Information sharing has also been identified as a dimension of smartness (Gil-Garcia et al. 2019), which can be used across agency boundaries and create complex information-based platforms. Furthermore, information sharing in urban planning information-based methods (Dang et al. 2006) has for long become common practices in planning and design processes, also allowing for participatory mechanisms or opportunities. Similarly, in operations such as for information system security (Lee et al. 2000) and supply chain networks (Li 2002), we see an extensive range of benefits from information-sharing and information technologies (Sanders and Henderson 2013). Thus, we see a growing opportunity for similar methods to be adapted and utilised for other sectors and (urban) systems. The central idea is to enhance the health–sustainability–peace (HSP) nexus in cities and communities.

In this chapter, we delve into information-sharing benefits as part of the peacebuilding mechanism during need. While we trust in the idea that 'perspective matters' (Chen and Sakamoto 2013), we also consider information sharing to be beneficial to decision-making processes and crisis management plans (Schraagen et al. 2010; Uitdewilligen and Waller 2018; Treurniet and Wolbers 2021). More importantly, we note the impact of good information-sharing mechanisms on promoting governance, particularly from the social dimension. The issues of social crisis information (Chen et al. 2018) and security (Phillips Jr et al. 2002) are truly in creating dynamism, interactions, and networks of communications (Zhao et al. 2018), between people and people, people and governments, and people and official social platforms. During disaster responses, we also see information sharing and coordination as critical elements in research and practice (Bharosa et al. 2010), which also shows the importance of communication in multiple processes, such as management processes, decision-making processes, and partnership processes.

Closely related to the sustainable development goals (SDGs), we note the importance of multiple perspectives and their differences (Cheshmehzangi and Dawodu 2019a, b) and how those differences could be managed and utilised appropriately. In order to support collaborative information sharing (Bentley et al. 1995), we ought to recognise the value of partnerships for sustainability transformations (Horan 2019), allowing in particular cross-sectoral partnerships (Cheshmehzangi 2020a) for information sharing, agenda setting, etc. (Horan 2019). A key aspect is also related to transparency in information sharing (Lamming et al. 2001; Milton 2009; Hartzler et al. 2011; Rahimi and Abedi 2020), which again should be considered cautiously (Maier and Ottaviani 2009) to ensure conflicts are minimised, misinformation does not occur, and trust is promoted throughout a process. This could lead to better policy principles for information-based transparency (Dawes 2010), which could help "*formulating a policy framework for public transparency initiatives*" (ibid), overcome some of the information-sharing issues and cross-boundary challenges, etc. Some of these approaches are extremely useful for organisational transparency enhancement, improving societal well-being and social relations with the governments.

This chapter may be the starting point for the social dimension of the HSP nexus; we also aim to highlight matters of information transparency, media management (and not control), information-sharing opportunities, and public initiatives. These are all essential to educate people as directly involved stakeholders for decision-making processes. In disaster management scenarios, we see information sharing as a noble method of facilitating response teams with accurate data and information (Waring et al. 2018). This also suggests the opportunity to use various information-based technologies in disaster relief (Bjerge et al. 2016) and have a better disaster response through information-sharing systems (Usuda et al. 2017). With a pinch of salt, we note matters of coordination and communication that could be enhanced through disaster response and disaster management processes. In this chapter, we further expand on these discussions and highlight the importance of information sharing from the social sustainability perspective.

12.2 Reflecting on How Cities Reacted During the Pandemic

We noted the importance of media management during the early phases of the COVID-19 pandemic (Cheshmehzangi 2020b), not from the control perspective but from good partnerships for promoting trust and transparency. Some other research also highlighted how enhanced media management could effectively manage COVID-19 stress (Zhao et al. 2020). Later, we highlighted effective methods of media coverage (Su et al. 2021), ensuring that related information is accurate. By minimising the negative impacts of media coverage, existing scholarly research suggests the role of communication, mainly through social media (Yu et al. 2020). However, the approach of crisis coordination cannot simply happen through one social media (Li et al. 2020) as it could cause further anxiety and uncertainty, leading to potential conflicts and mistrust. While social media is appraised for its capability for rapid knowledge dissemination (Chan et al. 2020), our concern remains on accurate and timely information-sharing methods or processes that are accurate and minimise negative consequences. Therefore, media exposure could be both positive and negative, depending on how it is managed (Cheshmehzangi 2020b) and how it is then amplified when and where needed (Garfin et al. 2020). As part of crisis management practices, mismanagement is often fatal (Ruiu 2020), showing that media management is also the main part of how information is narrated, shared, and ultimately disseminated.

During the early stage of the COVID-19, cities and communities were left with uncertainties, not knowing the depth and extent of the crisis. Thus, we see much of the coordination and communication occurred at a much larger scale, often at national level(s) or through international organisations and platforms. This uncertainty also caused some immediate issues for cities to cope with the immediate conflicts, misinformation, and management. Thus, we witnessed a large body of unverified information sharing (Laato et al. 2020), and any information could have disseminated correct or incorrect knowledge. To better understand publication relations, Thelwall and Thelwall (2020) suggest "*information dissemination and consensus building in all major forms, including on social media*". We note again the importance of transparency (Rahimi and Abedi 2020) and the accuracy of the information that must only be used for verified knowledge and better social exchanges (Xia et al. 2021). In this regard, we see emerging ICT-mediated platforms for information-sharing systems, such as the use of blockchain (Song et al. 2020), the development of data sets (Alqurashi et al. 2020; Zaman et al. 2020), and other ICT-based approaches for effective management scenarios (Strielkowski et al. 2021), and education (Naresh 2020).

In most cases, the higher-level information was directly shared at the city and community levels without considering contextual factors. This, in fact, could cause further misinformation or conflicts, making people unaware of the local situations and status. Information sharing at the city or community level requires social and official platforms, ensuring that trust and transparency are practised well, and information is disseminated correctly to all. In this regard, we express the fact that official media and its management in sharing accurate information have been truly important in developing a good relationship with people. When used as propaganda news, some information should be carefully assessed to ensure the right information and correct timing of dissemination. Some cities opt to have universal data-sharing standards (Allam and Joes 2020), which were quite helpful in terms of benefiting urban health and other aspects related to urban system operations. However, some cities do have specific characteristics, and their information would be more context-specific.

Everyday use of urban mobility information (Aloi et al. 2020) allowed many to reconsider their mobility behaviours. Some of these led to further anxiety in the use of public services (Cheshmehzangi 2020c). Still, some allowed firms to openly share information (Chesbrough 2020) and create a sharing economy activities platform (Hossain 2021). More importantly, we witnessed progressive approaches to consider spatial analysis to quickly verify social inequalities (Maroko et al. 2020), mainly using the available information and sharing the knowledge with a wider network of stakeholders. Some of these information-sharing and communication mechanisms helped to verify the needs for quality of life, primary health services for vulnerable groups, and find temporary and longer-term solutions to urban problems (Guida and Carpentieri 2021). The information-sharing systems also enabled many useful socio-economic predictors in detecting COVID-19 cases across various neighbourhoods and communities in cities like New York City (Whittle and Diaz-Artiles 2020). Some experimental cases also helped using sharing spatio-temporal information on identifying changing patterns of specific urban systems (Hu et al. 2021), enabling and enforcing various measures (Gupta et al. 2020), and create platforms where mobility data could help faster-tracking processes (Lamb et al. 2021). We see a growing number of information-sharing opportunities in all cases, emerging as experimentations and potential lessons (Chokshi and Katz 2020) for future pandemics. More importantly, such use of information sharing should be two sided, also benefitting the general public through better availability and transparent accessibility.

12.3 Suggestions Against the Common Flaws

As mentioned in previous chapters, cities are the places of experimentation and innovative interventions. Nonetheless, we ought to be cautious regarding information-sharing mechanisms, ensuring transparency, accuracy, and availability of information for all. Also, we need to provide information that is not misused or abused to create false propaganda or media coverage. We also should be aware of the mental health impacts of media coverage (Su et al. 2021), ensuring that information is shared in the right manner, i.e. with careful attention to the right channels and time of dissemination. In doing so, we are able to bring back people as main actors (Cheshmehzangi 2020a), not just through media channels but also through other information-sharing systems.

Information sharing is the main pillar of positive peace, also indicating the importance of trust and transparency in maintaining peace in society. In order to achieve such a level of trust and transparency, we ought to manage and disseminate data/information and not control what belongs to people. This data management aspect is critical to enhancing the resilience of cities and communities, identified as a dimension in the urban resilience framework against pandemics (Cheshmehzangi 2020a). To summarise accurately, we note the importance of information sharing in achieving HSP nexus in cities and communities. This could be done as an approach to enhance social sustainability and societal well-being.

The lessons from the COVID-19 pandemic on information sharing are threefold. The first is our suggestion for information sharing to the general public, allowing for trustbuilding through transparent procedures. The second is information sharing between firms and governmental units to create collaborative platforms where social demands could be maintained, participation could happen, and society could nourish. The third is information sharing through multiple media, which also includes social media. This requires media management mechanisms, allowing for an accurate and adequate level of information to all stakeholders. In achieving these, we are able to move towards smart-resilient models of cities and communities and, as such, to achieve a better HSP nexus from the social sustainability perspective.

12.4 A Summary

Our goal here is to ensure information sharing is explored from the social sustainability perspective. Some of the existing research studies suggest "*integrated platform models for national and international data-sharing and management*" (Gao et al. 2020), which are proposed as part of developing pandemic information-sharing directories. Thus, we see new international platforms creating directories for research on COVID-19 literature, such as the one established by the World Health Organization (WHO). However, we note that information sharing is more than just information shared between researchers or officials in different regions. In this chapter, we highlighted other methods of information sharing, such as media coverage opportunities, social platforms (e.g. social media), and mobility data. Researchers, practitioners, and officials should take extra caution when dealing with information sharing. Their role is to ensure ethical issues are kept safe, information is secure, and information sharing is done for positive purposes.

As part of information sharing in peace-building and post-conflict situations, we note the importance of trust and transparency. We also highlight there are two sides of the coin when it comes to such issues, suggesting growing cases of cyberattacks, information theft, information misuse, etc. As Lu et al. (2020) suggest, we also need to consider the mediating role of information sharing to allow people's contribution and direct involvement in information sharing and healthy use. In order to achieve a good HSP nexus in cities and communities, we recommend having local and official platforms where data can be exchanged and disseminated in a healthy manner. Examples of community smart city laboratories, such as those in cities like Amsterdam, are suitable experimental models that allow people to take part in information-sharing platforms or systems.

In sum, we highlight the importance of information sharing to manage information and data better and allow for healthy and secure exchanges. In recent years, we see big data and information-based platforms have become more fashionable than before. But we also argue that the state of the art is more than just creating complex systems and more towards collaborative opportunities that could break the barriers across governmental units, firms/practices, multiple sectors, etc. In doing so, we are able to create people-oriented information-sharing centres or platforms where society is placed at the heart of decision-making and information-sharing processes. In the next chapter, we delve into other people-related factors from the social sustainability perspective.

References

Allam Z, Jones DS (2020) On the coronavirus (COVID-19) outbreak and the smart city network: universal data sharing standards coupled with artificial intelligence (AI) to benefit urban health monitoring and management. In: Healthcare, vol. 8, No. 1. Multidisciplinary Digital Publishing Institute, p 46

Aloi A, Alonso B, Benavente J, Cordera R, Echániz E, González F, Ladisa C, Lezama-Romanelli R, López-Parra Á, Mazzei V, Perrucci L (2020) Effects of the COVID-19 lockdown on urban mobility: empirical evidence from the city of Santander (Spain). Sustainability 12(9):3870

Alqurashi S, Alhindi A, Alanazi E (2020) Large arabic twitter dataset on covid-19. *arXiv preprint* arXiv: 2004.04315. Available from: https://arxiv.org/abs/2004.04315

Anderson CC, Peace TP (2003) The antitrust risks of information sharing. Franchise LJ 23:17

Bentley R, Horstmann T, Sikkel K, Trevor J (1995) Supporting collaborative information sharing with the World Wide Web: The BSCW shared workspace system. In: Proceedings of the 4th International WWW Conference, vol. 1. IEEE Computer Society Press, Los Alamitos, CA, pp 63–74

Bharosa N, Lee J, Janssen M (2010) Challenges and obstacles in sharing and coordinating information during multi-agency disaster response: Propositions from field exercises. Inf Syst Front 12(1):49–65

Bjerge B, Clark N, Fisker P, Raju E (2016) Technology and information sharing in disaster relief. PloS One 11 (9):e0161783

Bormann NC, Cederman LE, Gates S, Graham BA, Hug S, Strøm KW, Wucherpfennig J (2019) Power sharing: Institutions, behavior, and peace. Am J Polit Sci 63(1):84–100

Butler JK Jr (1999) Trust expectations, information sharing, climate of trust, and negotiation effectiveness and efficiency. Group Org Manage 24(2):217–238

Cammett M, Malesky E (2012) Power sharing in postconflict societies: implications for peace and governance. J Conflict Resolut 56(6):982–1016

Chan AK, Nickson CP, Rudolph JW, Lee A, Joynt GM (2020) Social media for rapid knowledge dissemination: early experience from the COVID-19 pandemic. Association of Anaesthetists, https://doi.org/10.1111/anae.15057. Available from: https://associationofanaesthetists-publications.onlinelibrary.wiley.com/doi/full/10 .1111/anae.15057?af=R

Chen Y, Liang C, Cai D (2018) Understanding WeChat users' behavior of sharing social crisis information. Int J Hum-Comput Interact 34(4):356–366

Chen R, Sakamoto Y (2013) Perspective matters: sharing of crisis information in social media. In: 2013 46th Hawaii international conference on system sciences. IEEE, pp 2033–2041

Chesbrough H (2020) To recover faster from Covid-19, open up: managerial implications from an open innovation perspective. Ind Mark Manage 88:410–413

Cheshmehzangi A (2020a) The City in Need: Urban resilience and city management during disruptive disease outbreak events. Springer, Singapore

Cheshmehzangi A, Dawodu A (2019a) Sustainable urban development in the age of climate change. In: Cheshmehzangi A, Dawodu A (eds) Sustainable urban development in the age of climate change—people: the cure of curse. Palgrave Macmillan, Singapore, pp 157–182

Cheshmehzangi A, Dawodu A (2019b) The review of sustainable development goals (SDGs): people, perspective and planning. In: Cheshmehzangi A, Dawodu A (eds) Sustainable urban development in the age of climate change—people: the cure of curse, pp 133–156. Palgrave Macmillan, Singapore

Cheshmehzangi A (2020b) Responsiveness through city management. In: Cheshmehzangi A (eds) The city in need: urban resilience and city management during disruptive disease outbreak events. Springer, Singapore, pp 105–135

Cheshmehzangi A (2020c) COVID-19 and household energy implications: what are the main impacts on energy use? Heliyon 6(10):e05202

Chokshi DA, Katz MH (2020) Emerging lessons from COVID-19 response in New York City. JAMA 323 (20):1996–1997

Dang A, Wu L, Shi H, Han H (2006) Study on sharing digital urban planning information based on WebGIS. In: Geoinformatics 2006: geospatial information technology, vol. 6421. International Society for Optics and Photonics, p 64211V

Dawes SS (2010) Stewardship and usefulness: Policy principles for information-based transparency. Gov Inf Q 27(4):377–383

Gao F, Tao L, Huang YM, Shu Z (2020) Management and data sharing of COVID-19 pandemic information, Biopreservation Biobanking 18(6). https://doi.org/10.1089/bio.2020.0134

Garfin DR, Silver RC, Holman EA (2020) The novel coronavirus (COVID-2019) outbreak: amplification of public health consequences by media exposure. Health Psychol

Gates S, Graham BA, Lupu Y, Strand H, Strøm KW (2016) Power sharing, protection, and peace. J Politics 78(2):512–526

Gil-Garcia JR, Pardo TA, De Tuya M (2019) Information sharing as a dimension of smartness: understanding benefits and challenges in two megacities. Urban Aff Rev 1078087419843190

Guida C, Carpentieri G (2021). Quality of life in the urban environment and primary health services for the elderly during the Covid-19 pandemic: an application to the city of Milan (Italy). Cities 110:103038

Guo C, Li D, Zhang G, Zhai M (2018) Real-time path planning in urban area via vanet-assisted traffic information sharing. IEEE Trans Veh Technol 67 (7):5635–5649

Gupta M, Abdelsalam M, Mittal S (2020) Enabling and enforcing social distancing measures using smart city and its infrastructures: a COVID-19 Use case. arXiv preprint arXiv:2004.09246

Hartzell C, Hoddie M (2003) Institutionalizing peace: power sharing and post-civil war conflict management. Am J Polit Sci 47(2):318–332

Hartzler A, Skeels MM, Mukai M, Powell C, Klasnja P, Pratt W (2011) Sharing is caring, but not error free: transparency of granular controls for sharing personal health information in social networks. In: AMIA annual symposium proceedings, vol. 2011. American Medical Informatics Association, p 559

Horan D (2019) A new approach to partnerships for SDG transformations. Sustainability 11(18):4947

Hossain M (2021) The effect of the Covid-19 on sharing economy activities. J Cleaner Prod 280:124782

Hu S, Xiong C, Liu Z, Zhang L (2021) Examining spatiotemporal changing patterns of bike-sharing usage during COVID-19 pandemic. J Transp Geogr 91:102997

Ilcan S, Phillips L (2006) Governing peace: global rationalities of security and UNESCO's culture of peace campaign. Anthropologica 59–71

Kerschbaum F, Strüker J, Koslowski T (2011) Confidential information-sharing for automated sustainability benchmarks. Available from: https://aisel.aisnet.org/icis2011/proceedings/breakthroughideas/4/

Khan M, Hussain M, Papastathopoulos A, Manikas I (2018) Trust, information sharing and uncertainty: an empirical investigation into their impact on

sustainability in service supply chains in the United Arab Emirates. Sustain Dev 26(6):870–878
Kim J (2018) Social dimension of sustainability: from community to social capital. J Glob Scholars Market Sci 28(2):175–181
Laato S, Islam AN, Islam MN, Whelan E (2020) What drives unverified information sharing and cyberchondria during the COVID-19 pandemic? Eur J Inf Syst 29(3):288–305
Lai X, Tao Y, Wang F, Zou Z (2019) Sustainability investment in maritime supply chain with risk behavior and information sharing. Int J Prod Econ 218:16–29
Lamb MR, Kandula S, Shaman J (2021) Differential COVID-19 case positivity in New York City neighborhoods: socioeconomic factors and mobility. Influenza Other Respir Viruses 15(2):209–217
Lamming RC, Caldwell ND, Harrison DA, Phillips W (2001) Transparency in supply relationships: concept and practice. J Suppl Chain Manage 37(3):4–10
Lee HL, So KC, Tang CS (2000) The value of information sharing in a two-level supply chain. Manage Sci 46(5):626–643
Li L (2002) Information sharing in a supply chain with horizontal competition. Manage Sci 48(9):1196–1212
Li Y, Chandra Y, Kapucu N (2020) Crisis coordination and the role of social media in response to COVID-19 in Wuhan, China. Am Rev Publ Adm 50(6–7):698–705
Liburd J, Edwards DC (2010) The future of sustainability. Understanding the sustainable development of tourism. Published by Goodfellow Publishers Ltd., pp 225–237, Available from: https://opus.lib.uts.edu.au/handle/10453/16845
Liu GJ, Fu EJ, Wang YJ, Zhang KF, Han BP, Arrowsmith C (2007) A framework of environmental modelling and information sharing for urban air pollution control and management. J China Univ Min Technol 17(2):172–178
Lu L, Liu J, Yuan YC, Burns KS, Lu E, Li D (2020) Source trust and COVID-19 information sharing: the mediating roles of emotions and beliefs about sharing. Health Educ Behav. https://doi.org/10.1177/1090198120984760
Maier N, Ottaviani M (2009) Information sharing in common agency: when is transparency good? J Eur Econ Assoc 7(1):162–187
Maroko AR, Nash D, Pavilonis BT (2020) Covid-19 and Inequity: a comparative spatial analysis of New York City and Chicago hot spots. J Urban Health 97(4):461–470
Milton CL (2009) Information sharing: transparency, nursing ethics, and practice implications with electronic medical records. Nurs Sci Q 22(3):214–219
Naresh R (2020) Education after COVID-19 crisis based on ICT tools. UGC CARE J 31(37):464–468
Ott J, Hyytiä E, Lassila P, Vaegs T, Kangasharju J (2011) Floating content: Information sharing in urban areas. In: 2011 IEEE international conference on pervasive computing and communications (PerCom). IEEE, pp 136–146
Özer Ö, Zheng Y, Chen KY (2011) Trust in forecast information sharing. Manage Sci 57(6):1111–1137
Pearce JM (2012) Open source research in sustainability. Sustain J Rec 5(4):238–243
Phillips Jr CE, Ting TC, Demurjian SA (2002) Information sharing and security in dynamic coalitions. In: Proceedings of the seventh ACM symposium on access control models and technologies. pp 87–96
Porat R, Halperin E, Bar-Tal D (2015) The effect of sociopsychological barriers on the processing of new information about peace opportunities. J Conflict Resolut 59(1):93–119
Rahimi F, Abadi ATB (2020) Transparency and information sharing could help abate the COVID-19 pandemic. Infect Control Hosp Epidemiol 41(11):1366–1367
Ruiu ML (2020) Mismanagement of Covid-19: lessons learned from Italy. J Risk Res 23(7–8):1007–1020
Sanders CB, Henderson S (2013) Police 'empires' and information technologies: uncovering material and organisational barriers to information sharing in Canadian police services. Polic Soc 23(2):243–260
Schraagen JM, Veld MHIT, De Koning L (2010) Information sharing during crisis management in hierarchical versus network teams. J Contingencies crisis Manage 18(2):117–127
Shackelford SJ (2011) In search of cyber peace: A response to the cybersecurity act of 2012. Stan. L. Rev.Online, 64, 106. Available from: https://heinonline.org/HOL/Page?collection=journals&handle=hein.journals/slro64&id=108&men_tab=srchresults
Song J, Gu T, Feng X, Ge Y, Mohapatra P (2020) Blockchain meets COVID-19: a framework for contact information sharing and risk notification system. arXiv preprint arXiv:2007.10529
Strielkowski W, Firsova I, Lukashenko I, Raudeliūnienė J, Tvaronavičienė M (2021) Effective management of energy consumption during the COVID-19 pandemic: the role of ICT solutions. Energies 14(4):893
Su Z, McDonnell D, Wen J, Kozak M, Abbas J, Šegalo S, Li X, Ahmad J, Cheshmehzangi A, Cai Y, Yang L (2021) Mental health consequences of COVID-19 media coverage: the need for effective crisis communication practices. Glob Health 17(1):1–8
Thelwall M, Thelwall S (2020) Retweeting for COVID-19: consensus building, information sharing, dissent, and lockdown life. arXiv preprint arXiv:2004.02793, 10
Treurniet W, Wolbers J (2021) Codifying a crisis: progressing from information sharing to distributed decision-making. J Contingencies Crisis Manage 29(1):23–35
Uitdewilligen S, Waller MJ (2018) Information sharing and decision-making in multidisciplinary crisis management teams. J Organ Behav 39(6):731–748
Usuda Y, Hanashima M, Sato R, Sano H (2017) Effects and issues of information sharing system for disaster response. J Disaster Res 12(5):1002–1014

Vecchio G, Tricarico L (2019) "May the Force move you": roles and actors of information sharing devices in urban mobility. Cities 88:261–268

Waring S, Alison L, Carter G, Barrett-Pink C, Humann M, Swan L, Zilinsky T (2018) Information sharing in interteam responses to disaster. J Occup Organ Psychol 91(3):591–619

Whittle RS, Diaz-Artiles A (2020) An ecological study of socioeconomic predictors in detection of COVID-19 cases across neighborhoods in New York City. BMC Med 18(1):1–17

Xia J, Wu T, Zhou L (2021) Sharing of verified information about covid-19 on social Network sites: a social exchange theory perspective. Int J Environ Res Public Health 18(3):1260

Yu M, Li Z, Yu Z, He J, Zhou J (2020) Communication related health crisis on social media: a case of COVID-19 outbreak. Current Issues Tourism 1–7

Zaman A, Islam MN, Zaki T, Hossain MS (2020). ICT intervention in the containment of the pandemic spread of COVID-19: an exploratory study. arXiv preprint arXiv:2004.09888

Zhao X, Zhan M, Wong CW (2018) Segmenting and understanding publics in a social media information sharing network: an interactional and dynamic approach. Int J Strateg Commun 12(1):25–45

Zhao Y, An Y, Tan X, Li X (2020) Mental health and its influencing factors among self-isolating ordinary citizens during the beginning epidemic of COVID-19. J Loss Trauma 25(6–7):580–593

Human Capital and Equity: When the City Needs Its People

13

Our biggest problem is "poverty"—poverty in anything.

—Ali Cheshmehzangi, 2013

13.1 A Brief Exploration: Human Capital and Equity in People-Centric Cities and Communities

The topic of "human capital peace" is highly important in maintaining a good relationship between human capital and conflicts (Francis 2009; Namen et al. 2020). The healthy development of human capital has economic and social consequences, which are also part of the governance institutions supporting efficient production (Gilson and Roe 1999). The growing international flow of human capital, which has increased from decades ago (Grubel and Scott 1966), puts pressure on transferring services, peacebuilding programmes, and long-term sustainable development strategies. In peacebuilding, in particular, social capital plays a central part in "*social reintegration initiatives*" (Smoljan 2003), ensuring that human capital development and education are part of community building and societal enhancement exercises.

Some of the existing literature points out progressions from physical to human capital accumulation, particularly from the inequality perspective as part of development (Galor and Moav 2004). Some of these studies suggest further human capital investments (Rodrigeuz and Sanches 2012) towards prosperity and pro-growth entrepreneurship directions (Naudé 2007). A cross-country analysis conducted by Kodila-Tedika (2012) suggests a range of human capital variables for peacebuilding processes, some that are directly related to positive peace in all conditions. Despite its social sustainability importance, some scholars also look into physical and human capital accumulation related to peace development and economic growth plans (Santhirasegaram 2008; Bjørnskov 2012). In this regard, we see growing interest in research and policy development to include human capital as part of the sustainable social development of countries and regions. For cities, the situation is no different but perhaps less complicated.

Furthermore, social capital is identified as the primary mechanism in creating human capital (Coleman 1988). Nonetheless, the need for institutional development of human capital development (Acemoglu et al. 2014) cannot be separated from social sustainability perspectives. Hence, we also see societal education positions that help to verify better human capital development, inequality issues, and equity in communities and living/working environments. From an overarching perspective of "inclusive growth" (Raheem et al. 2018, we see opportunities for

A. Cheshmehzangi, *Urban Health, Sustainability, and Peace in the Day the World Stopped*, Sustainable Development Goals Series, https://doi.org/10.1007/978-981-16-4888-5_13

human capital development and those that are embedded in objectives of Millennium Development Goals (MDGs) and Sustainable Development Goals (SDGs). An example of such is methods of achieving health equity in society, proposing to be a catalyst for specific SDGs (Buzeti et al. 2020). This also reflects on finding adequate resources and mobilising them for health and human capital development (Kaboré et al. 2018). In a way, we can delve into subsequent human capital intervention, which includes "*productivity, equity, sustainability, and empowerment*" and towards poverty reduction measures (Ukeje et al. 2020). These are also rightly correlated with education-related SDG targets (Friedman et al. 2020), and more precisely, on health and education for building a healthier future from a human capital perspective (Bundy et al. 2020). We also note the role of human capital by having people involved during disaster management processes. City-specific research suggests human capital's role as part of developing networks of relationships (Simon and Nardinelli 1996, 2002). Some of these studies question the growth and institutional development of cities and regions from the perspective of human capital stock (Storper 2010), while the other cover divergence of human capital levels across cities (Berry and Glaeser 2005). Studies that focus on building community capital (Hancock 2001) usually focus on the social values of human progress and people partnership. They look into related aspects, such as production, economic life, and social sustainability (Arora et al. 2000).

Cities are recognised as human capital for migration (Benneworth and Herbst 2015) and attracting points for human capital progress. We see in the urbanising contexts, people play a significant part in urbanisation processes (Cheshmehzangi 2016), including the consideration of social capital development. Our investment in human capital is an important part of city development plans and strategies (Schultz 1961), which enable us to have further agendas on creativity and urban growth (Storper and Scott 2009). This chapter aims to explore two key aspects of human capital and equity, believing that the two come together in developing a sound people-centric city and/or community. We focus on how human capital and equity are important in achieving health–sustainability–peace (HSP) nexus, particularly during the time of need. Here, we reflect on the COVID-19 pandemic situation and discuss more on human capital and equity matters from the social sustainability perspective.

13.2 Reflecting on How Cities Reacted During the Pandemic

One of the evident cases of study on the topic relates to the impact of human capital efficiency (HCE) on equity funds (Yarovaya et al. 2021). The performance of such matter relates to human capital improvement plans and towards resilience enhancement in extreme periods such as the COVID-19 pandemic (ibid). We see the relationship between human capital and socio-economic attributes of society, enabling opportunities for efficiency enhancement (Mirza et al. 2020). By exploring the Human Capital Index (HCI), we are also able to see various measures, such as those related to benchmark critical components of human capital (World Bank 2020), particularly in the time of COVID-19. The existing examples suggest strategic human resource (HR) management (Cheshmehzangi 2020a, 2020b; Collings et al. 2021a), which of course, requires robust HR practices and human capital resources to start with, as the least of requirements. In the case of corporate resilience and response during the COVID-19, we see the importance of human capital and supply in maintaining economic operations (Cheema-Fox et al. 2020) and social development processes (Morgan and Trinh 2021). In the current time of social and economic crises, the poor communities are affected the most. Based on Morgan and Trinh's report of ASEAN countries (2021), we see specific socio-economic policies have been provided for social assistance of households such as in countries like Cambodia and Indonesia, social security fees have been revised in Myanmar, individual support is given for soft loan and tax relief, and social protection measures are adjusted in the Philippines. These are all excellent examples of utilising equity and fund for vulnerable communities, also assuring social services security and support.

Other cases reveal special human capital repurposing approaches, which push for the development of action plans for mitigating disparities (Peek et al. 2021). The importance of health equity and social capital is recognised as key aspects of disparities between households (Cheshmehzangi 2021) and communities in combatting the pandemic (Wang and Tang 2020). In countries like China, we see a growth of social capital, allowing for better human resource management, temporary solutions, and voluntary arrangements. However, the social equity factors may differ from context to context (Wright and Merritt 2020), also suggesting directions for better public administration in multinational countries like the USA. In achieving this, Wright and Merritt (2020) note the importance of prioritising social equity in crafting policy solutions and to better deal with a global health crisis. They also reflect on the earlier statement of Johnson and Svara (2015) that suggest the role of social equity in developing public policy, saying that it encompasses the:

> Fair, just and equitable management of all institutions serving the public directly or by contract, and the fair and equitable distribution of public services and the implementation of public policy and the commitment to promote fairness, justice, and equity in the formation of public policy.

Regarding how social equity has progressed, Guy and McCandless (2012) argue how it progressed from a philosophical to structural and administrative concern (see Wright and Merritt 2020). Thus, we note the importance of social equity in addressing healthcare inequality, social security, social services, etc. On health equity, we notice the importance of available human capital and workforces, suggesting better management for social policy development and recovery mechanisms (Lynch 2020). In this regard, some cities have genuinely struggled in maintaining good equitable support to those in need or had very little time to prepare against the adversities and after-effects of the lockdown measures, etc.

The other side of the human capital and equity debate focuses on the availability of human resources in cities and communities. As essential assets to society (Cheshmehzangi 2020a), people played an active part in managing and containing the outbreak. From human resources perspective, we note the importance of sectoral and cross-sectoral availability of essential workforces. The importance of human capital regarding having dedicated teams and support provisions (Wei et al. 2020) is highlighted as important in dealing with emotional and psychological situations caused by the COVID-19 pandemic. In their arguments related to human resources, Collings et al. (2021b) discuss sustainable performance in various ways, and mainly "*through the alignment of people and purpose and balancing the sort ad long term objectives*". Their viewpoint is valid regarding making HR leaders and capacity essential factors in driving the disaster management scenarios like the COVID-19 health emergency. At the city or community level, we see specific demographic determinants that show a wide range of issues regarding human resources, social capital, and equity. Thus, actions to mitigate disparities have been minimal initially and then become more apparent throughout multiple phases of the pandemic (Peek et al. 2021). A noteworthy factor is an interaction and collaboration between various human resources, which would be better managed if human capital were nurtured in the long run. Moreover, the intensity of human capital and knowledge could be playing a major part in enhancing urban agglomeration integration (Zheng and Du 2020) and opportunities for human capital development.

13.3 Suggestions Against the Common Flaws

During the COVID-19 pandemic, we witnessed the surge in inequalities, indicating issues related to lack of equity and resource availability to people. As evidenced by Cheshmehzangi

(2020a), city resilience would also depend on employee resilience and human capital (McCartney et al. 2021). The equity factors are also direct enablers in regard to better social capital and the social sustainability of cities. Thus, cities with better preparedness and better human capital management (such as human resource management) have been more successful in allocating people or essential workforces for specific sectors, providing support between sectors, and have adaptive plans in managing the pandemic (Cheshmehzangi 2020a). From the social sustainability perspective, we note the importance of equity in society, allowing for better preparedness and faster and smoother recovery processes. In contexts where equity is more widespread, we also see faster recovery processes after the containment of the disease at the city level. In this regard, we could partly reflect on particular institutional structures and governance, which embrace people-oriented and equitable approaches in human capital development. The existing situation also highlights shortfalls and shortage of human capital accumulation, which also affects human capital allocation and human resource management efficiency. We then recognise the nexus between equity and human capital, which we believe is important in ensuring social sustainability and creating a good relationship between people and city-level governments.

In addition, we see a growing need to prioritise health equity and social determinants of health, which is suggested nicely by Takian et al. (2020). Their study focuses on the relationship between social context and specific exposure results at the individual level, also allowing for a healthy process of policy development from the social consequences of ill health (ibid). Nonetheless, we ought to consider that such social factors may be influenced substantially by other factors, such as burdens from political sanctions (Raoofi et al. 2020), lower level social capitals and human resources (Cheshmehzangi 2020a), and other social networks that may be existent but not necessarily functional.

Lastly, we stress further human capital investment at the city and regional level, ensuring healthy labour mobility, social capital development, and equitable society structures. We note that human resource management is a major part of this specific factor, which should be understood as a two-way system. This means people should play a part in decision-making processes, and governments should enhance social capital development practices. We also stress the role of human capital accumulation, higher education levels, and provision of necessary training for communities. These are important to ensure communities are well-prepared for any adaptive measures and could deal with disaster management situations more promptly. During the COVID-19 pandemic, lessons learnt are mainly related to health equity and social capital deficiencies, which again could be dealt with through adequate and long-term investment plans. Furthermore, we notice the importance of a human capital pipeline that could help social capital development. In doing so, we are able to further invest in social capital enhancement processes, ensuring to attract human capital to be ready for specific situations or emergencies like pandemics.

13.4 A Summary

As discussed throughout this chapter, the two factors of "human capital" and "equity" are closely correlated, and, in some ways, are inseparable. The inequality issues (of any time) often come from a lack of equity in society, which over time could be amplified and widespread. We see such impacts were more visible in the contexts where societal disparities are significant and more perceptible. The mechanism for robust human capital development could lead us towards better social capital enhancement measures and opportunities. We could invest more in (societal) education, training, and preparedness in the form of action plans. This follows the same logic with compulsory military services practiced and/or adopted in some countries/contexts. In such countries, we see strong institutional mechanisms and support to develop a capacity of people and train them for the time of need, i.e. a war or conflict situation. The same approach should be considered in regard to training the general public and human capital in healthcare services, security and safety services, etc. We do not see the reason why such education and training cannot be widespread, and we hope human capital development could become the backbone of preparing cities for future emergencies like pandemics or even natural disasters.

To summarise briefly, we note the value of human capital in social capital development and vice versa (Coleman 1988). This is regarded to be relevant in HR management and the allocation of trained and educated people in specific workforces, such as in essential services. Bringing people back at the time of need is not something new, as we can see similar practices in wartime situations. Hence, the same applies to disaster situations, for us to prepare beforehand and react reflectively during such events. We stress again approaches that lead to better social capital investment, ensuring equity and social sustainability are central to community resilience enhancement. In the next chapter, we partly shift from the two covered social sustainability factors to other social sustainability indicators, such as social life in cities and communities. In doing so, we focus on public and social life and their role in the living environments, such as in public realms of cities and communities.

References

Acemoglu D, Gallego FA, Robinson JA (2014) Institutions, human capital, and development. Annu Rev Econ 6(1):875–912

Arora A, Florida R, Gates GJ, Kamlet M (2000) Human capital, quality of place, and location. Carnagie Mellon University, Pittsburgh, PA

Benneworth P, Herbst M (2015) The city as a focus for human capital migration: Towards a dynamic analysisof university human capital contributions. Eur Plan Stud 23(3):452–474

Berry CR, Glaeser EL (2005) The divergence of human capital levels across cities. Pap Reg Sci 84(3):407–444

Bjørnskov C (2012) How does social trust affect economic growth? South Econ J 78(4):1–21

Bundy D, Burbano C, Lloyd-Evans E, Sorgho G, Lavadenz F, Adam Z, Drake L, Rowland-Jones S, Piot P (2020) A commentary on the interagency symposium: 'building a healthier future: a human capital perspective on health and education', presented at the European congress of tropical medicine and international health, Liverpool, UK, October 2019. Int Health 12(4):235

Buzeti T, Madureira Lima J, Yang L, Brown C (2020) Leaving no one behind: health equity as a catalyst for the sustainable development goals. Eur J Public Health 30(Supplement_1):i24-i27

Cheema-Fox A, LaPerla BR, Serafeim G, Wang HS (2020) Corporate resilience and response during COVID-19. Available at SSRN 3578167

Cheshmehzangi A (2016) China's New-Type Urbanisation Plan (NUP) and the foreseeing challenges for decarbonization of cities: a review. Energy Procedia 104:146–152

Cheshmehzangi A (2020a) The city in need: Urban resilience and city management during disruptive disease outbreak events. Springer, Singapore

Cheshmehzangi A (2020b) Responsiveness through city management. In: Cheshmehzangi A (eds) The City in Need: Urban resilience and city management during disruptive disease outbreak events. Springer, Singapore, pp 105–135

Cheshmehzangi A (2021) Vulnerability of the UK's BAME communities during COVID-19: The review of public health and socio-economic inequalities. J Hum Behav Soc Environ 1–18

Coleman JS (1988) Social capital in the creation of human capital. Am J Sociol 94:S95–S120

Collings DG, McMackin J, Nyberg A J, Wright PM (2021a) Strategic human resource management and COVID-19: emerging challenges and research opportunities. J Manag Stud. Available from: https://www.ncbi.nlm.nih.gov/pmc/articles/pmc8013558/

Collings DG, Nyberg AJ, Wright PM, McMackin J (2021b) Leading through paradox in a COVID-19 world: Human resources comes of age. Hum Resour Manage J. Available from: https://onlinelibrary.wiley.com/doi/full/https://doi.org/10.1111/1748-8583.12343

Francis AM (2009) The Human capital peace: development and international conflict. Def Peace Econ 20 (5):395–411

Friedman J, York H, Graetz N, Woyczynski L, Whisnant J, Hay SI, Gakidou E (2020) Measuring and forecasting progress towards the education-related SDG targets. Nature 580(7805):636–639

Galor O, Moav O (2004) From physical to human capital accumulation: Inequality and the process of development. Rev Econ Stud 71(4):1001–1026

Gilson RJ, Roe MJ (1999) Lifetime employment: labor peace and the evolution of Japanese corporate governance. Colum L Rev 99:508

Grubel HB, Scott AD (1966) The international flow of human capital. Am Econ Rev 56(1/2):268–274

Guy ME, McCandless SA (2012) Social equity: its legacy, its promise. Public Adm Rev 72:S5–S13

Hancock T (2001) People, partnerships and human progress: building community capital. Health Promot Int 16(3):275–280

Johnson NJ, Svara JH (2015) Justice for all: promoting social equity in public administration: promoting social equity in public administration. Routledge, New York

Kaboré RMC, Solberg E, Gates M, Kim JY (2018) Financing the SDGs: mobilising and using domestic resources for health and human capital. The Lancet 392(10158):1605–1607

Kodila-Tedika O (2012) Determinants of peace: a cross-country analysis. Available from: https://mpra.ub.uni-muenchen.de/40812/

Lynch J (2020) Health equity, social policy, and promoting recovery from COVID-19. J Health Polit 45 (6):983–995

McCartney G, Pinto J, Liu M (2021) City resilience and recovery from COVID-19: The case of Macao. Cities 112:103130.

Mirza N, Hasnaoui JA, Naqvi B, Rizvi SKA (2020) The impact of human capital efficiency on Latin American mutual funds during Covid-19 outbreak. Swiss J Econ Stat 156(1):1–7

Morgan PJ, Trinh LQ (2021) Impacts of COVID-19 on households in ASEAN countries and their implications for human capital development. ADBI Working Paper 1226, Asian Development Bank Institute, Tokyo. Available from: https://www.adb.org/publications/impacts-covid-19-households-asean-countries

Namen O, Prem M, Vargas JF (2020) The human capital peace dividend. Available at SSRN 3424863

Naudé W (2007) Peace, prosperity and pro-growth entrepreneurship (No. 2007/02). WIDER Discussion Paper

Peek ME, Simons RA, Parker WF, Ansell DA, Rogers SO, Edmonds BT (2021) COVID-19 among African Americans: an action plan for mitigating disparities. Am J Publ Health (0):e1–e7

Raheem ID, Isah KO, Adedeji AA (2018) Inclusive growth, human capital development and natural resource rent in SSA. Econ Chang Restruct 51 (1):29–48

Raoofi A, Takian A, AkbariSari A, Olyaeemanesh A, Haghighi H, Aarabi M (2020) COVID-19 pandemic and comparative health policy learning in Iran. Arch Iran Med March 23(4):220–234. https://doi.org/10.34172/aim.2020.02

Rodriguez C, Sanchez F (2012) Armed conflict exposure, human capital investments, and child labor: evidence from Colombia. Def Peace Econ 23(2):161–184

Santhirasegaram S (2008) Peace and economic growth in developing countries: pooled data cross–country empirical study. In: International conference on applied economics–ICOAE 2008, pp 807–814

Schultz TW (1961) Investment in human capital: reply. Am Econ Rev 51(5):1035–1039

Simon CJ, Nardinelli C (1996) The talk of the town: Human capital, information, and the growth of English cities, 1861 to 1961. Explor Econ Hist 33(3):384–413

Simon CJ, Nardinelli C (2002) Human capital and the rise of American cities, 1900–1990. Reg Sci Urban Econ 32(1):59–96

Smoljan J (2003) The relationship between peace building and development. Confl Secur Dev 3(2):233–250

Storper M (2010) Why does a city grow? Specialisation, human capital or institutions? Urban Stud 47 (10):2027–2050

Storper M, Scott AJ (2009) Rethinking human capital, creativity and urban growth. J Econ Geogr 9(2):147–167

Takian A, Kiani MM, Khanjankhani K (2020) COVID-19 and the need to prioritize health equity and social determinants of health. Int J Publ Health 65:521–23. Available from: https://link.springer.com/content/pdf/https://doi.org/10.1007/s00038-020-01398-z.pdf

Ukeje IO, Ogbulu U, Amaefula VC (2020) Human capital intervention and poverty reduction. Global encyclopedia of public administration, public policy, and governance. Springer, Berlin

Wang Z, Tang K (2020) Combating COVID-19: health equity matters. Nat Med 26(4):458–458

Wei E, Segall J, Villanueva Y, Dang LB, Gasca VI, Gonzalez MP, Roman M, Mendez-Justiniano I, Cohen AG, Cho HJ (2020) Coping with trauma, celebrating life: reinventing patient and staff support during the COVID-19 pandemic: commentary describes how New York City Health+ Hospitals found ways to provide emotional and psychological support for patients, families, and staff during the COVID-19 pandemic. Health Aff 39(9):1597–1600

World Bank (2020) The human capital index 2020 update: human capital in the time of COVID-19. Available from: https://elibrary.worldbank.org/doi/abs/https://doi.org/10.1596/34432

Wright JE, Merritt CC (2020) Social equity and COVID-19: the case of African Americans. Public Adm Rev 80(5):820–826

Yarovaya L, Mirza N, Abaidi J, Hasnaoui A (2021) Human capital efficiency and equity funds' performance during the COVID-19 pandemic. Int Rev Econ Financ 71:584–591

Zheng S, Du R (2020) How does urban agglomeration integration promote entrepreneurship in China? Evidence from regional human capital spillovers and market integration. Cities 97:102529

14 Social and Public Life During Disruptive Times: A Public Realm Perspective

Alone, we may think big; but together, we can act big.
—Ali Cheshmehzangi, 2017

14.1 A Brief Exploration: Public Life, Social Life, and Public Realms of Cities

When it comes to social sustainability debates, public life is either neglected or is little discussed in the research. In fact, for us, it is important to understand public and social life as part of how cities could become healthier and more sustainable. In doing so, we are able to promote peace through quality public life, which would also determine the quality of life (QoL) and other well-being measures in cities and communities. Studies that focus on peace engineering normally reflect on such factors and address social sustainability factors, such as QoL and societal well-being. Expanded from earlier discussions of the previous chapter, we also note the importance of social equity and its relation to public life quality. Social equity is indeed an essential part of social sustainability performance, including a variety of social and public life measures.

In studies that focus on two factors of unity and diversity in public life, we see growing interest in democratic and social sustainability debates (Parker 2003). Some of these studies focus on issues of public services and social protection policies (Horton 2015), while others include broader discussions on public investment effects on social services and towards equality enhancement measures (Ilkkaracan et al. 2015). In all cases, we note the central position of social inclusivity, which is also utilised to promote Sustainable Development Goals (SDGs). However, when it comes to public life, we see a much larger scope of research, such as those related to public safety matters, social connectedness (Abraham 2021), and community empowerment through social protection policies (Thomas et al. 2019). The combined economic and public life factors are discussed more closely (Hoff et al. 2019) in achieving SDGs and considering relevant social determinants of Sustainable Development Plans. Nevertheless, when it comes to the actual public life in cities and communities, scholarly work neglects the importance of public realms and everyday places, particularly from the social sustainability perspective.

In places of conflict, we often see the role of public realms for social unrest and peacebuilding processes. They are the hot spots of where social life and public life meet and where people interact, participate in social activities, and (directly and indirectly) become part of social values. As part of peace culture debates (Boulding

A. Cheshmehzangi, *Urban Health, Sustainability, and Peace in the Day the World Stopped*, Sustainable Development Goals Series, https://doi.org/10.1007/978-981-16-4888-5_14

2002), we also note the importance of everyday life and social conditions of peace (Bahr 1973). The changing social dynamics of any society also determine progressions in keeping the public peace (Hindle 2002), produce social order (Lewis 2016), and maintain peace and stability in public life. According to Yarnall et al. (2021), more than one-quarter of the world population lives in countries where

> *…sustainable development outcomes are affected by fragility, conflict, and violence. Conflict is a critical development challenge that threatens the successful achievement of the Sustainable Development Goals. Business-as-usual practices that apply top-down approaches to development projects without considering the contextual dynamics are neither appropriate nor sufficient. To fully attain the Sustainable Development Goals, engineers must first recognize that development can only proceed when there is peace and stability, then must be prepared and willing to meet the complex challenges that threaten their long-term achievement.*

In this regard, we see continuous threats to peacebuilding practices, especially that social sustainability matters could be easily affected or fragile under specific political, economic, and even societal pressures. In cities and communities, we see such unrests commonly present in places of conflict, especially in everyday places, as part of public life, and within the livelihood of public realms. As the city's quintessential social territory (Lofland 2017), public realms are important with their proponents value peace (Luban 1994), which also resonate well with social sustainability and societal well-being in cities. They offer a range of infrastructural resources such as health services (Sennett 2020) and are places of education, peace, and culture (Gleeson 2006). As we delve into health–sustainability–peace nexus arguments, we reflect on some of the critical debates in urbanism. These include suggestions for healthy, sustainable, and vital neighbourhoods (Barton et al. 2003), which for long have been practices to enhance local health and meet global sustainability goals (Barton et al. 2006). In community and neighbourhood planning, we see the central role of social sustainability regarding good accessibility to public realms and the social infrastructure of cities (Karuppannan and Sivam 2011). We also correlate some of these viewpoints with other factors, such as public health (McMichael 2006; Cheshmehzangi 2020a), sustainability goals (Stefanescu, 2021), and offering co-benefits for achieving healthy living environments.

From the crossovers between public and political realms, we see the importance of human health in achieving (urban) sustainability (Scoones 2007). More importantly, we also note the significance of inclusivity, social engagement opportunities, and enhancing social responsibility in communities. Despite the many political and institutional constraints (Lodato and DiSalvo 2018), we note the importance of institutional structures and arrangement (Raven et al. 2019) not only in city transitions but also in defining the public realms. The livelihood formed from such social values determines what the HSP values of public realms could be. For instance, by reinventing the liveability of public realms (Cheshmehzangi 2015), we are able to augment the localities (Cheshmehzangi and Ornsby 2017) and suggest innovative ways to promote urban diversity and all-inclusive approaches to community-oriented planning (Cheshmehzangi and Li 2020). Thus, we highlight public realms as important social territories or nodes, through which public health could be enhanced, social sustainability could be achieved, and peace could be nurtured.

This chapter explores more than just sustainability perception of public realms and delves into further debates on how they affect cities and communities' public life and social life. We zoom into the specific factor of public life, exploring in particular overarching research areas of the public realm and social life. As cities and communities react to situations like pandemics, their social life becomes more vulnerable and yet remains the backbone of society. Thus, we aim to identify various aspects of social and public life in achieving urban health, sustainability, and peace.

14.2 Reflecting on How Cities Reacted During the Pandemic

The positive role of the public realm during public health emergencies has been appraised to enhance public health and community well-being (Wray et al. 2020). Despite the global challenges of utilising and accessing public places and recreational spaces (ibid), we note the importance of social life through an array of social networks, public sphere, and public realms (Hampton et al. 2010). Nonetheless, when it came to actual public realm use, issues of equity and access were questioned during the lockdown period of the COVID-19 pandemic (Cheshmehzangi 2020b; Lusk et al. 2021). While experimental and qualitative research highlights the effectiveness of green spaces in enhancing public health, we question availability and accessibility issues that may not be so widespread. Again, in cases where mental health effects were more substantial, we note the large-scale impacts of longer-term lockdown measures from the pandemic, which resulted in a lack of outdoor exercises, socialising events, etc. Accessibility to outdoor spaces enables better social opportunities even in smaller physical environments (Barron and Emmet 2021) and should also be protected when it comes to public place management (Cheshmehzangi 2020c, d). In these studies, we explore more adaptive measures for public realm management and monitory rather than control (Cheshmehzangi 2020d). Thus, we note the role of social life to have remained as part of public activities, social engagement, and urban relations that are important for the social sustainability of communities and cities.

In some cases, we also witnessed the fast-growing emergence of new technologies for further restructuring opportunities and dealing with mobility pattern behaviours in public places (Chen et al. 2020). Specific cases of higher-level surveillance and location-based mobility checks grew much faster in the boundary of social environments. As Hanson (2020) argues, "*the issue of surveillance in the public realm has come to the fore in this current health crisis*... [and]...*privacy has been supplanted by the needs of the collective good*". Surveillance and similar monitory tools appear to grow faster in the name of safety and security enhancement. Nonetheless, we may need to consider privacy and public good more cautiously (Hanson 2020) and ensure human connection could happen in various forms for benefitting public health (Cooley 2020). A significant factor is the vulnerability of the public realm, such as in cities like Tel Aviv (Lovencchio et al. 2020), which means that complex situations (e.g. climatic conditions, political issues, etc.) in the urban environments could lead to larger implications for recovery processes. Furthermore, we note a wide range of inequality issues even in public spaces of cities and communities (Honey-Rosés et al. 2020), which again reflects on issues that indicate how vulnerable groups became more vulnerable in the course of pandemic progression (Cheshmehzangi 2020c).

Globally, the images of empty streets, public realms, and social hubs of cities reminded us of the fragility of social attributes in our contemporary society. We have nurtured individualisation processes for a long time, and communities have become less selfless than ever. From public health enhancement to adaptive plans (Cheshmehzangi 2020c), we recognise the role of public realms more than just physical empty environments. The availability of and accessibility to public realms are basic needs of societal well-being, ensuring the most negligible impact on people's physical and mental health. Such considerations would allow us to have interventions for different social interactions and create adaptive spaces for different uses, i.e. for everyday uses, emergency or essential uses that may be most needed during the time of need, etc. From sheltering places to supporting essential services, we see a wide range of public realm uses, particularly from the public life perspective. The use of the public realm for emergency units, supply delivery places, community checkpoints, etc., are suitable examples that show the importance of outdoor and indoor public environments in maintaining daily operations and have emergency services/support. To maintain a social life, one must reflect on adaptive measures and utilise them for temporary transformations, reuse and reprogramming purposes, and social interventions. The latter is vital to enhance the health and peace of the general public, allowing for healthy livelihood while keeping the public safe from the pandemic.

14.3 Suggestions Against the Common Flaws

Scholars like Acuto (2020) suggest urban governance factors that help us better manage urban life (see Acuto et al. 2020). These are also seen to be more relevant for a more holistic understanding of the socio-spatial implications of COVID-19 measures (Salama 2020). Such implications could be important in evaluating symbiotic matters, such as social control, public health impacts, social life continuity, and social connectivity. For instance, in our earlier discussions, we reflect on the early closure of public places in regards to faster management and containment processes (Cheshmehzangi et al. 2021). Nevertheless, keeping the public realms safe should remain a priority, which should not be overlooked when investigating social life measures and practices.

In addition, we point out many socio-cultural differences and diversities (or diversity factors) that may exist in different locales. Thus, it is important to consider such differences with added attentiveness. In this regard, we ought to maintain societal livelihood, but without risking public health. To do so, adaptive planning and adaptive measures are highly recommended (Cheshmehzangi 2020d). In the form of adaptive strategies, such planning and measures help us to sustain health and at the same time achieve an optimal social life within and in between public realms.

The decline of public realms needs to be looked at more cautiously. This is not necessarily related to the ongoing pandemic, but as general thinking that applies to all times. We note the evident values of public life and social life in our public realms, which are, in fact, the basis of our livelihood, quality of life, and societal well-being. Undoubtedly, the implications would also be on achieving peace in communities, where we see more opportunities for social cohesion, social connectedness, and social values. These strengthen our social responsibility, care about our living/working environments, and help us appreciate public life from a social sustainability perspective.

Furthermore, we note how public realms should maintain a livelihood while remaining safe environments during pandemics. The decline in people's trust of public services and amenities and public realms could lead to further implications on growing individualisation and negative societal structures. For long, we have already sabotaged some of the core social values of our traditional livelihoods, such as co-housing living environments and intergenerational living. We cannot afford to lose anything further than we have lost in terms of our core social values. Therefore, public realms should be protected to ensure they remain as social hubs in cities, particularly in the digital age. We, as humans, are social animals and therefore should be able to safeguard what is believed to be our social values, social connectivity, and social norms.

To go against the common flaws, we have to understand the adaptive attribute of our public realms. We have to know their social co-benefits and the not-so-hidden social connectedness that determine the social values of future generations. In the current age of digitalisation and digitisation (of almost everything), we should know the values of social interactions and connections that may fade away in the near future. For public realms, we have to consider temporary arrangements, repurposing procedures of the social environments, and move towards enhancing the social infrastructure of cities and communities. On social infrastructure, in particular, we expressed only a little, but we know its values for the empowerment of "public" and "public life". Thus, we urge to reconsider the transformative transitions of our public realms, ensuring that they remain in place for our society's sustainability, safety, and vitality. By achieving this goal, peace is maintained within the placeness of our public life.

14.4 A Summary

This chapter serves as an exploration of public life and social life through the capacity of public realms. We could learn a lot from successful examples of adaptive planning in public places across many cities around the globe. From early adaptive measures to support emergency and healthcare services to later spatial rearrangements and repurposing strategies, we note the value of public realms more than just physical environments. The social values that are embedded in our public realms should not be forgotten, particularly if we aim to enhance our public health, social infrastructure, and social services.

As we close our investigation of various aspects under the social sustainability dimension, we highlight the overlapping discussions between social and other five dimensions. In particular, the obvious crossovers with economic and institutional factors are truly represented in achieving a sound health–sustainability–peace (HSP) nexus in our communities and cities. In the next part of the book, we delve into the institutional dimension, focusing on some of the essential factors that should be embedded in achieving urban health, longer-term sustainability, and peaceful living and working environments. The social dimension offers many normative and regulative conceptions, which would undoubtedly be beneficial to the enhancement of public health, peacebuilding practices, and sustainability transitions.

References

Abraham DB (2021) Local SDG Indicators for US Cities and Communities. In: Promoting the sustainable development goals in north american cities: case studies & best practices in the science of sustainability indicators, pp 147–155

Acuto M (2020) COVID-19: lessons for an urban (izing) world. One Earth 2(4):317–319

Acuto M, Larcom S, Keil R, Ghojeh M, Lindsay T, Camponeschi C, Parnell S (2020) Seeing COVID-19 through an urban lens. Nat Sustain 1–2

Bahr HE (1973) The politicising of everyday life: social conditions of peace. J Peace Res 10(1–2):37–49

Barron C, Emmet MJ (2021) Back gardens and friends: the impact of COVID-19 on children and adolescents use of, and access to, outdoor spaces. Ir Geogr 53 (2):173–177

Barton H, Grant M, Guise R (2003) Shaping neighbourhoods: a guide for health, sustainability and vitality. Taylor & Francis, London

Barton H, Grant M, Guise R (2006) Shaping neighbourhoods: for local health and global sustainability. Routledge, New York and London

Boulding E (2002) Peace culture. Compassionate 8. Available at: https://learningpartnership.org/sites/default/files/resources/pdfs/copanthology.pdf#page=12

Chen B, Marvin S, While A (2020) Containing COVID-19 in China: AI and the robotic restructuring of future cities. Dialogues Hum Geogr 10(2):238–241

Cheshmehzangi A (2015) The reinvention of liveability in public places: interaction mapping analysis of central Nottingham's improved walkability. J Hum Behav Soc Environ 25(5):426–440

Cheshmehzangi A (2020a) Reflection on early lessons for urban resilience and public health enhancement during the COVID-19. Health 12(10):1390

Cheshmehzangi, A. (2020b). Housing and health evaluation related to general comfort and indoor thermal comfort satisfaction during the COVID-19 lockdown. J Hum Behav Soc Environ 1–26

Cheshmehzangi A (2020c) The city in need: urban resilience and city management during disruptive disease outbreak events. Springer, Singapore

Cheshmehzangi A (2020d) 10 Adaptive measures for public places to face the COVID 19 pandemic outbreak. City Soc 32(2)

Cheshmehzangi A, Li HMA (2020) Innovation through urban diversity and achieving comprehensive sustainable urbanism from a community-oriented approach. Curr Urban Stud 8(02):222

Cheshmehzangi A, Ornsby P (2017) Augmented locality: the utilisation of urban screens in public places as new networks of the city. J Urban Regeneration Renewal 11(2):194–211

Cheshmehzangi A, Sedrez M, Ren J, Kong D, Shen Y, Bao S, Xu J, Su Z, Dawodu A (2021) The effect of mobility on the spread of COVID-19 in light of regional differences in the European Union. Sustainability 13(10):1169794

Cooley L (2020) Fostering human connection in the COVID-19 virtual health care realm. Nejm Catal Innovations Care Delivery. Available from: https://www.ncbi.nlm.nih.gov/pmc/articles/pmc7371327/

Gleeson B (2006) Desocializing space: the decline of the public realm in Western Sydney. Soc Cult Geogr 7 (1):19–34

Hampton KN, Livio O, Sessions Goulet L (2010) The social life of wireless urban spaces: internet use, social networks, and the public realm. J Commun 60(4):701–722

Hanson F (2020) Surveillance in the public realm in the post-Covid-19 world. Planning News 46(7):18–19

Hindle S (2002) The keeping of the public peace. In: Hindle S (ed) The state and social change in early modern England 1550–1640. Palgrave Macmillan, London, pp 94–115

Hoff H, Monjeau A, Gomez-Paredes J, Frank F, Rojo S, Malik A, Adams K (2019) Spillovers in SDG implementation. Available from: https://www.jstor.com/stable/resrep22973

Honey-Rosés J, Anguelovski I, Chireh VK, Daher C et al (2020) The impact of COVID-19 on public space: an early review of the emerging questions–design, perceptions and inequities. Cities Health 1–17

Horton R (2015) Offline: gender equality—the neglected SDG for health. The Lancet 386(10007):1928

Ilkkaracan I, Kim K, Kaya T (2015) The impact of public investment in social care services on employment, gender equality, and poverty: the Turkish case. Research Project Report, Istanbul Technical University Women's Studies Center in Science, Engineering and Technology and the Levy Economics Institute, in partnership with ILO and UNDP Turkey, and the UNDP and UN Women Regional Offices for Europe and Central Asia. Available from: https://www.kaum.itu.edu.tr/en/wp-content/uploads/2015/09/rapor-son-eng.pdf

Karuppannan S, Sivam A (2011) Social sustainability and neighbourhood design: an investigation of residents' satisfaction in Delhi. Local Environ 16(9):849–870

Lewis D (2016) Central Asia: Contested Peace. In: Richmond O, Pogodda S, Ramovic J (eds) The Palgrave handbook of disciplinary and regional approaches to peace. Palgrave Macmillan, London, pp 387–397

Lodato T, DiSalvo C (2018, August). Institutional constraints: the forms and limits of participatory design in the public realm. In: Proceedings of the 15th participatory design conference: full papers, vol 1. pp 1–12

Lofland LH (2017) The public realm: exploring the city's quintessential social territory. Routledge, New York and London

Lovecchio J, Basic G, Pawlowski T (2020) Urban heat, vulnerability, and the public realm: lessons from Tel Aviv-Yafo and implications for COVID-19 recovery. Smart Sustain Fair Cities 40:108–136

Luban D (1994) Settlements and the erosion of the public realm. Georgetown Law J 83:2619–2662. Available from: https://heinonline.org/HOL/Page?handle=hein.journals/glj83&div=74&g_sent=1&casa_token=

Lusk K, Einstein K, Glick D, Palmer M, Park S, Fox S (2021) Urban parks and the public realm: equity & access in post-COVID cities. Available from: https://open.bu.edu/handle/2144/42350

McMichael AJ (2006) Population health as the 'bottom line' of sustainability: a contemporary challenge for public health researchers. Eur J Public Health 16 (6):579–581

Parker WC (2003) Teaching democracy: unity and diversity in public life. Teachers College Press, part of Columbia University Press, New York

Raven R, Sengers F, Spaeth P, Xie L, Cheshmehzangi A, de Jong M (2019) Urban experimentation and institutional arrangements. Eur Plan Stud 27(2):258–281

Salama AM (2020) Coronavirus questions that will not go away: interrogating urban and socio-spatial implictions of COVID-19 measures. Emerald Open Res 2. Available from: https://www.ncbi.nlm.nih.gov/pmc/articles/PMC7219560/

Scoones I (2007) Sustainability. Dev Pract 17(4–5):589–596

Sennett R (2020) The public realm. In: Goldhill S (ed) Being urban: community, conflict and belonging in the middle east, part of planning, history, and environment series. Routledge, New York and London, pp 35–58

Stefanescu CA (2021) Sustainability reporting in the public realm—trends and patterns in knowledge development. Sustainability 13(8):4128

Thomas M, Carey E, Ice DB, Berfond J et al (2019) Women's economic empowerment: strengthening public and private sector impact through accountability and measurement (SDG 5). JICA Res Inst 2030

Wray A, Fleming J, Gilliland J (2020) The public realm during public health emergencies: exploring local level responses to the COVID-19 pandemic. Cities Health 1–4

Yarnall K, Olson M, Satiago I, Zelizer C (2021) Peace engineering as a pathway to the sustainable development goals. Technol Forecast Soc Change 168:120753

Part V

Sustainability and Peace: "Urban Health from the Institutional Dimension"

In the fifth part of studying sustainability and peace, we present three primary aspects representing urban health from the institutional dimension.

The first aspect is focused on an overarching topic of public rights, which is discussed as a supporting instrument for social and public services. From the institutional perspective, we look into this factor as part of social infrastructure development, justice and partnerships, and societal improvements. We delve into some of the common inequalities and suggest ways of institutional response. These points are discussed in Chap. 15, mainly focused on civil and human rights matters. The second aspect of this part is primarily related to the discourse of good relations with neighbours. As one of the main positive peace pillars, good relations with neighbours are explored through regional thinking and planning to further discuss the sustainability of cities beyond their borders. The focus here would be more on institutional structures that would affect infrastructural development, regional productions, regional regulations and policymaking procedures, and preserving peace in the process of pandemic events. This aspect is explored in detail in Chap. 16, where we aim to debate institutional governance and argue in favour of promoting good neighbour relations. Topics of institutional access, governance, and frameworks are discussed as part of the dynamic of regional production networks. The third aspect zooms into the importance of public networks because of sound institutional structures. In Chap. 17, we specifically explore the importance of public networks for a healthy society and discuss more on institutional support that is needed for peacebuilding processes.

Public Rights and Maintaining the City's Social and Public Services 15

Evolution is by choice.
—Ali Cheshmehzangi, 2019

15.1 A Brief Exploration: Public Rights and Achieving Healthy Public Services

Public or civil rights are important part of "*public rights of action premised upon administrative and institutional interests*" (Davis 2014). Some of the scholarly work on this topic focuses on public administration institutional choice in specific sectors (Horn and Horn 1995). Some argue about regulatory matters and political practices (Magarian 2002) encompassing institutional evolution (Buergenthal 1997), institutional analysis (Schmid 1972), institutional framework (Scully 1988), and institutional isomorphism (Frumkin and Galaskiewicz 2004). In land use and urban planning, we evaluate some of these aspects from the perspective of public planning (Alexander 2001) and participatory planning processes (Cheshmehzangi and Dawodu 2019a, b; Dawodu et al. 2019). However, when it comes to urban studies, much of the work on public rights is often limited to other factors such as property rights (Webster 2007) and legislation for public and private right change (Rose 1996). However, some studies also express concerns regarding racial inequalities (Epstein 2006; Sugrue 2006), class and rationality matters (Woodard 1988), and place-making processes (Carter 2014). Some of these are critical in achieving healthy public services, particularly during the time of calamities.

Giorgetta (2002) argues various points regarding the right to a healthy environment, human rights, and sustainable development. Such debates are particularly relevant to multiple factors such as the right to a healthy environment, including environmental rights, participatory democracy, and the right to public access (ibid). These are also important in achieving healthy public policy (Milio 2001), which could support related factors of public health design and planning (Frumkin et al. 2004), public health enhancement (Cheshmehzangi 2020a,2021; Su et al. 2021), and developing an adequate level of public health policy (Lindström and Eriksson 2009). In their recommendations for the concepts and form of healthy people, Fielding and Kumanyika (2009) reflect on some of the key aspects of public health issues, participation in the development of healthy people, and encouraging more institutional support. Therefore, public rights are essential in developing healthy policymaking procedures (MacNaughton 2015), empowering people (Mølhave and Krzyzanowski 2003), and enhancing healthy public policy through health equity measures (Richmond and

A. Cheshmehzangi, *Urban Health, Sustainability, and Peace in the Day the World Stopped*,
Sustainable Development Goals Series, https://doi.org/10.1007/978-981-16-4888-5_15

Cook 2016). Nonetheless, without institutional support, these could simply not happen. As it happens in many contexts, especially in the global south, we notice long-term issues of minimal institutional support or instruments in developing people-centric policies. The lack of institutional framework often adds to widespread pressures on cities and communities, making them more vulnerable during emergencies or other calamities. The right to a healthy city has been discussed in relation to various aspects of placemaking, policymaking, and manifestations of human rights as the common source for equity and equality improvements (Skinner and Masuda 2013). Some existing literature on healthy city debates (Corburn 2009, 2013; Leeuw and Simos 2017) truly reflects on some of the human well-being matters as part of civil/public rights, which is right in terms of mobilising financial and human resources (Harpham et al. 2001), and providing guidance to policies and policymakers for healthy cities (Cole et al. 2008). In this regard, we see growing interest on putting human right next to ideas of achieving healthy environment (Knox and Pejan 2018), which show the importance of "Declaration on the Human Dimension of Global Climate Change", as well as matters that relate to liveability, quality of life, and social/societal well-being.

This chapter is the starting chapter for debating institutional factors, allowing us to provide a broader picture of institutional factors concerning achieving urban health or healthy cities. This chapter focuses on public rights, which is recognised as a supporting instrument for social and public services. We discuss the topic in relation to the provision of social/public services (Cheshmehzangi 2020b) and towards social infrastructure development, justice and partnerships, and societal improvements. As part of creating social bridges, the social infrastructure helps to ensure better integration between different social groups, empowerment, and better services for local communities (Vaznonienė and Pakeltienė 2017). We also verify the importance of creativity and productivity, through which people could enjoy "*self-respect and guaranteed human rights*" (Gopalakrishna and Leelavathi 2011). In regard to the development of social standards (Likosky 2006), we see growing debates in management practices of local authorities (Frolova et al. 2016), which also include constitutional factors and determinants of social infrastructure (Eicher et al. 2018). Arguments on entrepreneurial social infrastructure, in particular, suggest ways of achieving human capital (Flora and Flora 1993; Flora et al. 1997), ensuring their rights are protected, and social infrastructure services are provided (Vaznonienė and Kiaušienė 2018) through adequate institutional support. Thus, this chapter delves into the debates of human rights mainly from the institutional perspective.

15.2 Reflecting on How Cities Reacted During the Pandemic

As shown already in many published studies reflecting on the COVID-19 impacts, we see the interdependency and magnitude of impacts concerning specific sectors (Baum and Hai 2020). Addressing human rights has been a central message of international organisations like the World Health Organization (WHO), whose director suggested "*all countries must strike a fine balance between protecting health, minimizing economic and social disruption, and respecting human rights*" (WHO 2020). Nonetheless, issues of human rights, especially regarding public health, are often quite complicated and require adequate health and human rights framework (Mykhalovskiy et al. 2020). The human rights of vulnerable groups were also seen as a significant challenge, questioning procedures, response and control plans, protocols, and safety measures (Schiariti 2020).

The international human rights law indicates that it "*guarantees everyone the right to the highest attainable standard of health and obligates governments to take steps to prevent threats to public health and to provide medical care to those who need it*" (Amon and Wurth 2020). Nevertheless, we see many examples of limits of human rights (Spadaro 2020), such as those related to human rights imperatives (Todd-

Gher and Shah 2020), public health matters, economic rights, and global solidarity (Sekalala et al. 2020). Thus, we not see the promised international protection and often witnessed lack of human rights at the smaller scales of regions, cities, and communities. Moreover, as the COVID-19 pandemic grew larger, there were many signals of many countries limiting their human rights (Dzehtsiarou 2020), even though it should be not be seen as an "*obstacle to effective governmental measures targeting the pandemic*" (ibid). However, due to our widespread inequality issues, we noticed a tangible widening gap in human rights practices of specific contexts.

When treated like a wartime situation, as happened during the course of the COVID-19 pandemic, human rights may become more vulnerable. Thus, we noted how cities became fragile in addressing the needs of vulnerable groups, as the inevitable decline in social services (Cheshmehzangi 2020b) led to further adversities and pressures on the poorer groups. Many urged to respect human rights amid the pandemic (Siviwe 2020) and highlighted "*a grossly inadequate public health and social care infrastructures for undervalued and powerless groups*" (Amadasun 2020). This means the situation poses "*legitimate concerns to social policy response and social work professionals*" (ibid). For instance, the lack of institutional support for inclusive approaches (Armitage and Nellums 2020) indicates concerns regarding generalised approaches of lockdown, quarantines, self-isolation, and sheltering-in-place. With eroded human rights, as highlighted by Openshaw and Travassos (2020), we see wide impacts on the impoverished populations and those marginalised groups like essential workers, disabled groups, the elderly, etc. We also see growing inequalities in mandatory quarantine measures, especially for the poor populations who face hefty costs and effects on their daily working operations. More to these viewpoints are weakening migrant protocols that continuously lack protection and adequate public support (Garrett 2020), also worsening human rights disasters and mental health consequences.

As evidence shows, the derogation to human rights during the COVID-19 pandemic (Lebret 2020) impacted larger-scale management and containment of the disease. Thus, it is vital to address the issues at a smaller scale, ensuring that collective rights are protected, and inclusive measures are considered. In this regard, we see growing legal implications such as for those related to public rights during self-isolation and quarantine (Coghlan et al. 2020a), which in a way is in place to protect a larger group than just the individuals. Hence, we see a growing importance of public rights rather than individual rights, which again may have a negative impact on growing disparities and inequalities between different social groups, genders, ages, etc.

In the time of the COVID-19 pandemic, there are cases of the erosion of human rights for the poor (Hotez et al. 2020), showing a large (and potentially a long-term) impact on social justice, food loss (and subsequent food security), food access and waste, etc. (Fleetwood 2020). We also see the adverse impacts on the progression of implementing the Sustainable Development Goals (SDGs), leading to weaker institutional structures and operations for public health. We see through evidence a much larger financial pressure on social and public services in cities, which again mean lower performance of such services at large.

15.3 Suggestions Against the Common Flaws

Through a human dignity perspective, Colombo (2020) argues issues of governmentalities, exploring the issues of public and political debate on various topics, such as biomedical and economical. We see neglect and sometimes inactions in maintaining fundamental human rights during the pandemic (Chimowa et al. 2020), which means the need for more careful and considerate measures in facing the economic recession (Bohoslavsky 2020) and declining human rights in and after the pandemic. In many cases, much of the actions had potentially undermined human rights, which also expanded on societal vulnerabilities such as in global value chains (Voss 2020). In cases where people tracing is a common practice, issues of endangering human rights (Coghlan et al. 2020b) may have been lesser than places where they face new moral and public health challenges. Hence, meaningful and holistic measures are recommended to "*recognize the breadth of the challenges to both economic, social, and cultural rights, and civil and political rights*" (Bennoune 2020). Thus, legal and social protection measures and policies are recommended (Neef 2020) to ensure that long-term recovery programmes are contextualised, swiftly adapted, and successfully implemented.

As the pandemic affects the vulnerable more severely, we note the importance of maintaining social and public services through robust institutional support (Cheshmehzangi 2020b). As countries and regions prepare themselves to tackle the longer-term adversities, cities and communities should ensure having the right and inclusive institutional structures for better support and immediate action plans. In the case of the occupied Palestinian territory, for instance, Moss and Majadle (2020) argue issues such as border closures and many barriers that have already weakened the movement of goods. In such contexts, or in war zones and places under heavy sanctions, the impacts are large scale, and human rights are generally neglected. Lack of medical support and treatment is just the basic rights that could be simply neglected in such locales. Shortfall and shortcomings in the city's essential systems could lead to devastating consequences, which for long has been ignored through nasty political lobbying and hidden bilateral agreements between countries. At the city and community level, such matters become more perceptible. Lack of public rights could become an obstacle against resilience enhancement, public health support, and better handling of the outbreaks. For the future, such conflicts should be dealt with more promptly with the rightful engagement of the international organisations, rather than the decisions of allies against the common enemy.

15.4 A Summary

Humanity, as it exists, is fragile already; the lack of institutional support to public rights could make our cities and communities even more delicate. The COVID-19 crisis has become much more than the initial health emergency (Cheshmehzangi 2020b). Over its progression, the pandemic has highlighted some of the flaws of our institutional structures and frameworks, showing large-scale corruption issues, human rights issues, etc. As cities and communities grow to become independent and self-sufficient while connected to global networks, they need to have a specific and contextualised set of public rights that could be supported by their local institutions and governments.

This chapter looks at some of the conflicting issues with human rights, which concern us about the future after the pandemic and how people mobility, migration protection, and other related factors could eventually evolve. By investing more in (local) institutions that need to respond to social needs and provide necessary public services, we could see better opportunities for the position of public rights than just the business-as-usual scenarios. This chapter covered only the tip of a very large iceberg, showing (only) what was commonly experienced or witnessed. If we ought to explore the issues in depth, we need to examine specific regimes, political status, institutional structures, and many other factors affecting public rights. In the next chapter, we continue on primary institutional factors; but, we will zoom out to the regional scale, where we believe the relations between cities and their neighbours, and/or communities and their neighbours, play a significant part in achieving a peaceful and sustainable society.

References

Alexander ER (2001) A transaction-cost theory of land use planning and development control: towards the institutional analysis of public planning. Town Plann Rev 45–75

Amadasun S (2020) COVID-19 palaver: ending rights violations of vulnerable groups in Africa. World Develop 134:105054. Available from: https://www.ncbi.nlm.nih.gov/pmc/articles/PMC7315991/#b0005

Amon JJ, Wurth M (2020) A virtual roundtable on COVID-19 and human rights with human rights watch researchers. Health Hum Rights 22(1):399

Armitage R, Nellums LB (2020) The COVID-19 response must be disability inclusive. Lancet Public Health 5 (5):e257

Baum T, Hai NTT (2020) Hospitality, tourism, human rights and the impact of COVID-19. Int J Contemp Hosp Manag 32(7):2397–2407. https://doi.org/10.1108/IJCHM-03-2020-0242

Bennoune K (2020) "Lest We Should Sleep": COVID-19 and Human Rights. Am J Int Law 114(4):666–676

Bohoslavsky JP (2020) COVID-19: urgent appeal for a human rights response to the economic recession. United Nations Human Rights Special Procedures. Geneva. Available from: https://www.yourhomeworksolutions.com/wp-content/uploads/edd/2020/09/66pgpaper5.pdf

Buergenthal T (1997) The normative and institutional evolution of international human rights. Hum Rights Q 19(4):703–723

Carter RL (2014) Valued lives in violent places: black urban placemaking at a civil rights memorial in New Orleans. City Soc 26(2):239–261

Cheshmehzangi A (2020) Reflection on early lessons for urban resilience and public health enhancement during the COVID-19. Health 12(10):1390

Cheshmehzangi A (2020) The city in need: urban resilience and city management during disruptive disease outbreak events. Springer, Singapore

Cheshmehzangi A (2021) Urban Health and Sustaining Peace during and after the COVID-19 Pandemic. J Infect Dis Prev Med 9(3) 1000219:1–3. Available from: https://www.longdom.org/open-access/urban-health-and-sustaining-peace-during-and-after-the-covid19-pandemic.pdf

Cheshmehzangi A, Dawodu A (2019a) Case study reviews: people, perspective and planning. In: Cheshmehzangi A, Dawodu A (eds) Sustainable urban development in the age of climate change—people: the cure or curse. Palgrave Macmillan, Singapore, pp 69–131

Cheshmehzangi A, Dawodu A (2019b) The review of sustainable development goals (SDGs): people, perspective and planning. In: Cheshmehzangi A, Dawodu A (eds) Sustainable urban development in the age of climate change—people: the cure or curse. Palgrave Macmillan, Singapore, pp 133–156

Chimowa T, Hall S, O'Hare B (2020) Public money creation to maintain fundamental human rights during the COVID-19 pandemic. Health Hum Rights 22 (1):395

Coghlan N, Archard D, Sipanoun P, Hayes T, Baharlo B (2020a) COVID-19: legal implications for critical care. Anaesthesia 75(11):1517–1528

Coghlan S, Cheong M, Coghlan B (2020b) Tracking, tracing, trust: contemplating mitigating the impact of COVID-19 through technological interventions. Interventions 213:6–8

Cole D, Lee-Smith D, Nasinyama G (eds) (2008) Healthy city harvests: generating evidence to guide policy on urban agriculture. Int Potato Center. Available from: http://www.fao.org/fileadmin/templates/FCIT/PDF/healthy_city_harvestes.pdf

Colombo E (2020) Human rights-inspired governmentality: COVID-19 through a human dignity perspective. Crit Sociol 0896920520971846. Available from: https://doi.org/10.1177/0896920520971846

Corburn J (2009) Toward the healthy city. Massachusetts Institute of Technology, Cambridge, Massachusetts. Available from: https://mitp-web2.mit.edu/sites/default/files/titles/content/9780262513074_sch_0001.pdf

Corburn J (2013) Healthy city planning: from neighbourhood to national health equity. Routledge, New York and London

Davis S (2014) Implied public rights of action. Columbia Law Rev 114:1–84

Dawodu A, Cheshmehzangi A, Williams A (2019) Expert-initiated integrated approach to the development of sustainability indicators for neighbourhood sustainability assessment tools: an African perspective. J Cleaner Prod 240:117759

de Leeuw E, Simos J (eds) (2017) Healthy Cities: the theory, policy, and practice of value-based urban planning. Springer-Verlag, New York

Dzehtsiarou K (2020) COVID-19 and the European convention on human rights. Strasbourg Observers 27. Available from: https://clr.iliauni.edu.ge/wp-content/uploads/2020/07/Kanstantsin-Dzehtsiarou-pp.53-58.pdf

Eicher TS, García-Peñalosa C, Kuenzel DJ (2018) Constitutional rules as determinants of social infrastructure. J Macroecon 57:182–209

Epstein KK (2006) A different view of urban schools: civil rights, critical race theory, and unexplored realities, vol 291. Peter Lang Publishing Inc., Bern

Fielding J, Kumanyika S (2009) Recommendations for the concepts and form of healthy people 2020. Am J Prev Med 37(3):255–257

Fleetwood J (2020) Social justice, food loss, and the sustainable development goals in the era of COVID-19. Sustainability 12(12):5027

Flora CB, Flora JL (1993) Entrepreneurial social infrastructure: a necessary ingredient. Ann Am Acad Pol Soc Sci 529(1):48–58

Flora JL, Sharp J, Flora C, Newlon B (1997) Entrepreneurial social infrastructure and locally initiated

economic development in the nonmetropolitan United States. Sociol Q 38(4):623–645
Frolova EV, Vinichenko MV, Kirillov AV, Rogach OV, Kabanova EE (2016) Development of social infrastructure in the management practices of local authorities: trends and factors. Int J Environ Sci Educ 11 (15):7421–7430
Frumkin H, Frank L, Frank LD, Jackson RJ (2004) Urban sprawl and public health: designing, planning, and building for healthy communities. Island Press
Frumkin P, Galaskiewicz J (2004) Institutional isomorphism and public sector organizations. J Public Adm Res Theor 14(3):283–307
Garrett TM (2020) COVID-19, wall building, and the effects on migrant protection protocols by the Trump administration: the spectacle of the worsening human rights disaster on the Mexico-US border. Administrative Theory & Praxis 42(2):240–248
Giorgetta S (2002) The right to a healthy environment, human rights and sustainable development. Int Environ Agreements 2(2):171–192
Gopalakrishna BV, Leelavathi DS (2011) Infrastructure and Human development in India: An Inter-state comparison. J Glob Econ 7(4):292–311
Harpham T, Burton S, Blue I (2001) Healthy city projects in developing countries: the first evaluation. Health Promot Int 16(2):111–125
Horn M, Horn MJ (1995) The political economy of public administration: Institutional choice in the public sector. Cambridge University Press
Hotez PJ, Huete-Perez JA, Bottazzi ME (2020) COVID-19 in the Americas and the erosion of human rights for the poor. PLOS Neglected Trop Dis 14(12):e0008954. https://doi.org/10.1371/journal.pntd.0008954
Knox JH, Pejan R (eds) (2018) The human right to a healthy environment. Cambridge University Press, Cambridge
Lebret A (2020) COVID-19 pandemic and derogation to human rights. J Law Biosci, 7(1), lsaa015
Likosky MB (2006) Law, infrastructure and human rights. Cambridge University Press, Cambridge
Lindström B, Eriksson M (2009) The salutogenic approach to the making of HiAP/healthy public policy: illustrated by a case study. Glob Health Promot 16(1):17–28
Magarian GP (2002) Regulating political parties under a public rights first amendment. William Mary Law Rev 44(5):1939–2062
MacNaughton G (2015) Human rights impact assessment: a method for healthy policymaking. Health Hum Rights J 17:63–75
Milio N (2001) Glossary: healthy public policy. J Epidemiol Community Health 55(9):622–623
Mølhave L, Krzyzanowski M (2003) The right to healthy indoor air: status by 2002. Indoor Air 13:50–53
Moss D, Majadle G (2020) Battling COVID-19 in the occupied Palestinian territory. Lancet Glob Health 8 (9):e1127–e1128
Mykhalovskiy E, Kazatchkine C, Foreman-Mackey A, McClelland A, Peck R, Hastings C, Elliott R (2020) Human rights, public health and COVID-19 in Canada. Can J Public Health 111(6):975–979
Neef A (2020) Legal and social protection for migrant farm workers: lessons from COVID-19. Agric Hum Values 37:641–642
Openshaw JJ, Travassos MA (2020) COVID-19, quarantines, sheltering-in-place, and human rights: the developing crisis. Am J Trop Med Hyg 103(2):578–580
Richmond CA, Cook C (2016) Creating conditions for Canadian aboriginal health equity: the promise of healthy public policy. Public Health Rev 37(1):1–16
Rose CM (1996) A dozen propositions on private property, public rights, and the new takings legislation. Wash Lee L Rev 53:265
Schiariti V (2020) The human rights of children with disabilities during health emergencies: the challenge of COVID-19. Dev Med Child Neurol 62(6):661
Schmid AA (1972) Analytical institutional economics: challenging problems in the economics of resources for a new environment. Am J Agr Econ 54(5):893–901
Scully GW (1988) The institutional framework and economic development. J Polit Econ 96(3):652–662
Sekalala S, Forman L, Habibi R, Meier BM (2020) Health and human rights are inextricably linked in the COVID-19 response. BMJ Global Health 5(9): e003359
Siviwe B (2020) South Africans urged to 'respect human rights' amid COVID-19 pandemic. Retrieved from: https://www.thesouthafrican.com/news/south-africans-urged-to-respect-human-rights-amid-covid-19-pandemic/
Skinner E, Masuda JR (2013) Right to a healthy city? Examining the relationship between urban space and health inequity by Aboriginal youth artist-activists in Winnipeg. Soc Sci Med 91:210–218
Spadaro A (2020) COVID-19: testing the limits of human rights. Eur J Risk Regul 11(2):317–325
Su Z, Wen J, McDonnell D, Goh E, Li X, Šegalo S, Xiang YT (2021) Vaccines are not yet a silver bullet: the imperative of continued communication about the importance of COVID-19 safety measures. Brain Behav Immunity-Health 12:100204
Sugrue TJ (2006) Affirmative action from below: civil rights, the building trades, and the politics of racial equality in the urban north, 1945–1969. Best Am History Essays 2006:231–261
Todd-Gher J, Shah PK (2020) Abortion in the context of COVID-19: a human rights imperative. Sexual Reprod Health Matters 28(1):1758394

Vaznonienė G, Kiaušienė I (2018) Social infrastructure services for promoting local community wellbeing in Lithuania. European countryside. Warsaw: De Gruyter 10(2)

Vaznonienė G, Pakeltienė R (2017) Methods for the assessment of rural social infrastructure needs. European countryside. Warsaw: De Gruyter, 9(3)

Voss H (2020) Implications of the COVID-19 pandemic for human rights and modern slavery vulnerabilities in global value chains. Trans Corporations J 27(2). Available from: https://ssrn.com/abstract=3692319

Webster C (2007) Property rights, public space and urban design. Town Plann Rev 81–101

Woodard MD (1988) Class, regionality, and leisure among urban black Americans: the post-civil rights era. J Leis Res 20(2):87–105

WHO (World Health Organization) (2020) Addressing human rights as key to the COVID-19 response. Available from: https://covid19-evidence.paho.org/handle/20.500.12663/1190

16 Good Relations with Neighbours: Sustainability of Cities Beyond Their Borders

We all have our differences, I agree! But, we need to learn how to understand the differences, how to respect them, and how to make a better use of them.

—Ali Cheshmehzangi, 2015

16.1 A Brief Exploration: Good Relations that Matter the Most

From the institutional perspective, good neighbourly relations are essential in terms of respecting human rights (Noutcheva 2015). In all contexts and across all sectors, public relations are essential in developing and maintaining good institutional arrangements (Cass et al. 2010) and better institutional governance. The effectiveness of institutional frameworks suggests opportunities for good relations with neighbours, one of the eight primary pillars of positive peace. In most cases, cross-border cooperation, especially with immediate neighbours, is not necessarily easy. Everyday peace is simply not possible when it comes to neighbour relations, especially in between countries of the same region. At the smaller scale of cities and/or communities, this matter could differ as we mainly have the obstacle of competitiveness and not major conflicts, such as those that may exist between countries or regions. However, there are cases of exceptions in places where segregated communities are the norm or have become the norm after years/decades of conflict. Examples of segregation could relate to land, resources, accessibilities, trade, religion, etc. In all cases, however, security matters play a major part in maintaining peace across communities, regardless of their long-term or temporary nature.

When it comes to good relations with neighbours, we often consider measures and practices related to peace and security. From the larger sub-regional and continental scales (Higashino 2004) to country-level and regional scales, we see growing attention to public diplomacy and public relations in order to strengthen good neighbourly relations. By learning from conflict-affected contexts, we see the importance of bottom-up approaches and local agencies (Mac Ginty 2014) more than just those international involvements that usually stay for long or leave the context with poorer neighbourly relations. In her arguments on the illusion of peace, Marks (2003) argues about territorial conflicts and bad relations that would shape because of such conflicts. We see similar challenges for some of the landlocked developing countries (Faye et al. 2004), where peace and stability are very much dependant on the immediate neighbours or transit countries across the borders. However, in places where good relations matter the most, we see opportunities to materialise long-term relations

A. Cheshmehzangi, *Urban Health, Sustainability, and Peace in the Day the World Stopped*,
Sustainable Development Goals Series, https://doi.org/10.1007/978-981-16-4888-5_16

that may serve as contextualised peacebuilding processes and for the benefit of multiple parties.

Furthermore, we could pursue stability or stable relationships through institutional governance and related policies that enable us to improve relations with neighbours (Chung 2009). Both differences and similarities between countries of the same region could create positive and negative effects on relations. Nonetheless, sub-regional and regional cooperation could determine better collaborative platforms where economic development and trade, sociocultural exchanges, environmental policies, and institutional development could happen. For instance, despite their historical differences, the US–Canada relations have been one of the good examples of good relations with neighbours. Essentially, the two sides have found more commonalities than those unnecessary differences.

Following Buzan's discourse on "peace, power, and security" (Buzan 1984), we see the role of neighbour relations to become more effective at the time of need. Creating a healthy ecosystem of relations is essential to any operations at any level, but in most cases, such relationships do not occur or are sometimes very fragile. The same applies to the smallest living environments, such as households, where inter-family disagreements could lead to eventual conflicts, longer-term effects on the whole family, and even separations. The same logic also applies to working environments, communities and neighbourhoods, cities, and regions. In all cases, we see the role of external relations as essential externalities for peacebuilding processes, ensuring that institutional governance and frameworks are in place, peace treaties are agreeable, and good neighbourly attitudes are practised.

Despite the interest in international relations and multilateral or bilateral relations, we mainly focus on regional relations and those that exist between cities of the same region. In doing so, we aim to highlight the effective role of regional planning and decision-making and elaborate on what means a good model of neighbour relations. As a primary positive peace pillar, it is essential to see how good relations could promote health and sustainability in communities and cities of the same region. Thus, in this chapter, we delve into an overarching discourse of good relations with neighbours, exploring more about regional thinking and planning to achieve sustainability in a region and beyond city borders. The focus here would be more on institutional structures that would affect infrastructural development, regional productions, regional regulations and policymaking procedures, and preserving peace in the process of pandemic events. We aim to debate institutional governance and argue in favour of promoting good neighbour relations. Therefore, it is relevant to discuss topics of institutional access, governance, and frameworks, as part of the dynamic of regional production networks.

16.2 Reflecting on How Cities Reacted During the Pandemic

More than any time, we are living in a globalised world. The globalising trend truly affects our self-sufficient plans at the smaller scale and enables us to work collaboratively with others to achieve larger goals, such as peacebuilding and sustainability target plans. During the COVID-19 pandemic, good neighbourly relations, robust international relations, and public diplomacy helped to respond more effectively and rapidly to regional and city-level situations. Although regional COVID-19 dynamics exist (Cheshmehzangi et al. 2021; Engbert et al. 2021), there are also signs of successful regional coordination (Cheshmehzangi 2020a; Easom et al. 2020; Pasin et al. 2020). In reality, other regional conditions could affect the spread of COVID-19, such as mobility patterns across the regions (Cheshmehzangi et al. 2021), regional climate condition (Iqbal et al. 2020), regional hygiene matters (Rovetta and Castaldo 2020), and regional tourism (Uğur and Akbıyık 2020). Thus, as verified in many analytical and modelling studies, multiple factors influence regional vulnerability to COVID-19 (Tahmasebi et al. 2020), which suggest the importance of intermittent regional strategies (Della Rossa et al. 2020) and a better action plan for regional differences (Hoekman et al. 2020). These factors directly relate to good relations with neighbours through diplomacy and unwavering relations from the institutional and peacebuilding perspectives.

Successful examples evidently showed regional cooperation under various circumstances, such as the global production network (Kimura et al. 2020) or immediate regional action plans (Cheshmehzangi 2020a). Also, we see positive institutional support in developing public health protection policy response (Shvetsova et al. 2020) or immediate policy design developed from state capacity and the regional institutional arrangements (Capono 2020). In many cases, especially in the case of China, we see the role of regional resilience (Gong et al. 2020), which indicates that previous institutional experiences led to the successful allocation of "*government support schemes, as well as regional industrial structures*" (ibid). Amid the immediate action plans to COVID-19, we note China's response showed strong cooperation at the regional level and the institutional support that was only derived from robust institutional structures and frameworks (Cheshmehzangi 2020b). In other cases, such as the first European epicentre during the first wave of the pandemic, Italy used its regional institutional autonomy to respond to shortfall and shortages in hospitals, essential and emergency services, healthcare systems, etc. (Garattini et al. 2020). A similar approach was seen in some of the African regional responses, which indicated the activation of regional plans that overlapped the existing regional institutional framework (Medinilla et al. 2020). Moreover, in cities like Seattle, we see successful models of divisional and institutional preparatory response (Parikh et al. 2020), ensuring adaptive measures are considered (Cheshmehzangi 2020b, c) while regional variabilities and institutional differences are understood prior to making action plans. Another example is the successful approach to having healthcare workers in regional referral hospitals in Northern Uganda (Amanya et al. 2021), which led to better coordination between urban and rural areas of the same region and more efficient use of institutional support at the regional level.

In between countries, sharing regional information has been quite impactful in creating regional action plans (Rossman et al. 2020). Such an approach allows better prevention and control measures, such as those immediately developed and implemented by the small Oceanian countries (Cheshmehzangi 2020a). Some examples, such as the Oceanian model, consider fostering regional economic prosperity despite the large-scale impacts on their individual economies. At a smaller scale, such as at the provincial or state scale, we see opportunities for regional institutional support, such as the development of regional regulations and public health measures (Cheshmehzangi 2020d). A good example that we could reflect on is regional food production, which indicates the dependency of urban areas on their surrounding outskirt and rural areas (Chen et al. 2020). This approach helps to enhance urban–rural relations and create a win-win scenario for the economic prosperity of rural areas and essential supplies, such as food, for the urban areas. The approach also led to some regional surveillance systems, such as those developed at the provincial/federal levels in countries like Australia, China, Germany, and Japan. In particular, the approach was helpful to consider a more manageable scale for productions, mobility patterns, healthcare support, etc.

16.3 Suggestions Against the Common Flaws

In our earlier arguments, we noted the importance of regional institutional involvement in developing contextualised and localised regulations and response action plans (Cheshmehzangi 2020e). This consideration helps boost collective institutional action during the pandemic (Wilson et al. 2020), focused on various aspects such as local and regional economic development. As suggested by Della Rossa et al. (2020), intermittent regional strategies develop a network model helping to alleviate or at least control the pandemic at a manageable scale. While regional variations should be taken into consideration (Bellizzi et al. 2020; Guzzi et al. 2020; Shim 2021), we note opportunities to reduce risk factors and increase regional cooperation. For instance, in regard to mitigation strategies, multi-institutional collaboration has proven to be effective in ensuring pandemic alleviation and better control (Lee et al. 2020).

From regional food production to regional energy sources and supplies, we see greater opportunities for health–sustainability–peace (HSP) creation with good neighbourly relations. As evidenced in the case of China during the management of the COVID-10 pandemic, regional measures and strategies were proven to be more effective than national-level plans. The early closure of provinces and regions helped to enhance regional resilience as well as productions that later supported the economic stability at the city and regional levels. It was mainly through regional strategies that the country managed to progress effectively in its first nationwide lockdown period, allowing for provinces and city clusters to work more effectively together. The same model was also seen in some other countries where regional boundaries are defined, and institutional frameworks are robust at the regional level. The two ought to come together to ensure the successful implementation of strategies at the regional level through good relations with neighbours and collective institutional support.

Our suggestions are divided into four categories. The first is related to a regional scale that includes several countries, ensuring regional development is a primary goal between countries of the same region. This is a model, which has been developing in regions like Oceania and South East Asia, and in models like the European Union. Nonetheless, this approach requires regional institutional frameworks that could support collective decision-making and support between the countries. The second is related to regional support in regard to economic development, which includes channels of economic networks and flows between countries or cities in the same region, allowing for better mobility and relations and sharing resources more effectively. The third is related to institutional access and regional governance aspects, including the development of relevant institutional frameworks and a better capacity and monitory of regional production networks. This approach helps to minimise differences and share prosperity between multiple stakeholders and multiple institutional networks. In doing so, cities of the same region could create a better mechanism to support one another with some collective core targets and shared values. The fourth is related to urban–rural relations as part of regional thinking, which is growing fast in countries where urban–rural balance is affected. Key aspects, such as production(s) and economic development, could nourish through such urban–rural relations. The approach is vital in maintaining good relations with neighbours and allowing institutional support for both urban and rural parts of the same region.

16.4 A Summary

The institutional perspective on any type of city-level experimentation would depend on specific institutional arrangements that could significantly differ between cities or the built environments (Raven et al. 2019). As evident throughout the COVID-19 pandemic, we see the role of collective institutional support or schemes in sustaining productions and economic networks. The same applies when it comes to the provision of healthcare support, emergency services, food and amenities supplies, and other essential services. At the regional level, we could see the importance of good neighbourly relations, regardless of scale. Nonetheless, regional variations and differences could turn into obstacles without collective and regional institutional frameworks. Thus, we urge relevant actors and stakeholders to consider such an approach in developing good relations, enhancing collective targets and goals, and implementing strategies at the regional level.

To summarise, we note the importance of this critical positive peace pillar, which is truly relevant to regional resilience factors when explored from the institutional perspective. To ensure a reasonable HSP nexus is achieved, we ought to consider collective measures and strategies beyond the city boundaries, enabling us to think about what is the best for a larger scale of region or sub-region. This chapter explored some of the primary matters related to regional thinking, which should provide us with some suggestions for future regional planning directions. These discussions will be followed in the next chapter, where we explore more about public networks, some that could be related to regions, but some that could be related to smaller scales of cities and communities.

References

Amanya SB, Nyeko R, Obura B, Acen J, Nabasirye C, Nakaziba R et al (2021) Knowledge and compliance with covid-19 infection prevention and control measures among health workers in regional referral hospitals in Northern Uganda: a cross-sectional online survey. F1000Research 10:136

Bellizzi S, Napodano CMP, Salaris P, Pichierri G, Sotgiu G (2020) Regional variation in trajectories of healthcare worker infections during the COVID-19 pandemic in Italy. Infect Control Hosp Epidemiol 41 (12):1472–1474

Buzan B (1984) Peace, power, and security: contending concepts in the study of International Relations. J Peace Res 21(2):109–125

Capano G (2020) Policy design and state capacity in the COVID-19 emergency in Italy: if you are not prepared for the (un) expected, you can be only what you already are. Policy Soc 39(3):326–344

Cass N, Walker G, Devine-Wright P (2010) Good neighbours, public relations and bribes: the politics and perceptions of community benefit provision in renewable energy development in the UK. J Environ Planning Policy Manage 12(3):255–275

Chen K, Zhang Y, Zhan Y, Fan S, Si W (2020, February 20). How China can address threats to food and nutrition security from the Coronavirus outbreak, professional article on IFPRI Blog. Available at: https://www.ifpri.org/blog/how-china-can-address-threats-food-and-neutrition-security-coronovirus-outbreak

Cheshmehzangi A (2020a) COVID-19 and the small commonwealth Oceania countries: promising regional co-ordination. The round Table 109(4):466–467

Cheshmehzangi A (2020b) Recommendations for 'The city in need'. In: Cheshmehzangi A (ed) The city in need: urban resilience and city management during disruptive disease outbreak events. Springer, Singapore, pp 285–304

Cheshmehzangi A (2020c) Reflection on Disruptions: managing the city in need, saving the city in need. In: Cheshmehzangi A (ed) The city in need: urban resilience and city management during disruptive disease outbreak events. Springer, Singapore, pp 137–283

Cheshmehzangi A (2020d) Reflection on early lessons for urban resilience and public health enhancement during the COVID-19. Health 12(10):1390

Cheshmehzangi A (2020e) The city in need: urban resilience and city management during disruptive disease outbreak events. Springer, Singapore

Cheshmehzangi A, Sedrez M, Ren J, Kong D, Shen Y, Bao S, Xu J, Su Z, Dawodu A (2021) The effect of mobility on the spread of COVID-19 in light of regional differences in the European Union. Sustainability 13(10):5395. https://doi.org/10.3390/su13105395

Chung CP (2009) The" Good Neighbour Policy" in the context of China's foreign relations. China Int J 7 (1):107–123

Della Rossa F, Salzano D, Di Meglio A, De Lellis F, Coraggio M, Calabrese C, Guarino A, Cardona-Rivera R, De Lellis P, Liuzza D, Iudice FL (2020) A network model of Italy shows that intermittent regional strategies can alleviate the COVID-19 epidemic. Nat Commun 11(1):1–9

Easom N, Moss P, Barlow G, Samson A, Taynton T, Adams K, Ivan M, Burns P, Gajee K, Eastick K, Lillie PJ (2020) Sixty-eight consecutive patients assessed for COVID-19 infection: experience from a UK regional infectious diseases unit. Influenza Other Respir Viruses 14(4):374–379

Engbert R, Rabe MM, Kliegl R, Reich S (2021) Sequential data assimilation of the stochastic SEIR epidemic model for regional COVID-19 dynamics. Bull Math Biol 83(1):1–16

Faye ML, McArthur JW, Sachs JD, Snow T (2004) The challenges facing landlocked developing countries. J Hum Dev 5(1):31–68

Garattini L, Zanetti M, Freemantle N (2020) The Italian NHS: what lessons to draw from COVID-19? Appl Health Econ Health Policy 18:463–466

Gong H, Hassink R, Tan J, Huang D (2020) Regional resilience in times of a pandemic crisis: the case of COVID-19 in China. Tijdschr Econ Soc Geogr 111 (3):497–512

Guzzi PH, Tradigo G, Veltri P (2020) Spatio-temporal resource mapping for intensive care units at regional level for COVID-19 emergency in Italy. Int J Environ Res Public Health 17(10):3344

Higashino A (2004) For the sake of 'peace and security'? The role of security in the European Union enlargement eastwards. Coop Confl 39(4):347–368

Hoekman LM, Smits MMV, Koolman X (2020) The Dutch COVID-19 approach: regional differences in a small country. Health Policy Technol 9(4):613–622

Iqbal MM, Abid I, Hussain S, Shahzad N, Waqas MS, Iqbal MJ (2020) The effects of regional climatic condition on the spread of COVID-19 at global scale. Sci Total Environ 739:140101

Kimura F, Thangavelu SM, Narjoko D, Findlay C (2020) Pandemic (COVID-19) policy, regional cooperation and the emerging global production network. Asian Econ J 34(1):3–27

Lee AK, Cho RH, Lau EH, Cheng HK, Wong EW, Ku PK et al (2020) Mitigation of head and neck cancer service disruption during COVID-19 in Hong Kong through telehealth and multi-institutional collaboration. Head Neck 42(7):1454–1459

Mac Ginty R (2014) Everyday peace: bottom-up and local agency in conflict-affected societies. Secur Dialogue 45(6), 548–564

Marks S (2003) The illusion of peace: international relations in Europe 1918–1933. Macmillan International Higher Education, London and New York

Medinilla A, Byiers B, Apiko P (2020) African regional responses to COVID-19 (No. 272). Discussion Paper

Noutcheva G (2015) Institutional governance of European neighbourhood policy in the wake of the Arab Spring. J Eur Integr 37(1):19–36

Parikh SR, Bly RA, Bonilla-Velez J, Dahl JP, Evans SS, Horn DL et al (2020) Pediatric otolaryngology divisional and institutional preparatory response at Seattle Children's Hospital after COVID-19 regional exposure. Otolaryngology-Head Neck Surg 162 (6):800–803

Pasin L, Sella N, Correale C, Boscolo A, Rosi P, Saia M, Mantoan D, Navalesi P (2020) Regional COVID-19 network for coordination of SARS-CoV-2 outbreak in Veneto, Italy. J Cardiothorac Vasc Anesth 34 (9):2341–2345

Raven R, Sengers F, Spaeth P, Xie L, Cheshmehzangi A, de Jong M (2019) Urban experimentation and institutional arrangements. Eur Plan Stud 27(2):258–281

Rossman H, Keshet A, Shilo S, Gavrieli A, Bauman T, Cohen O et al (2020) A framework for identifying regional outbreak and spread of COVID-19 from one-minute population-wide surveys. Nat Med 26(5):634–638

Rovetta A, Castaldo L (2020) The Impact of COVID-19 on Italian web users: a quantitative analysis of regional hygiene interest and emotional response. Cureus 12(9)

Shim E (2021) Regional variability in covid-19 case fatality rate in canada, february–december 2020. Int J Environ Res Public Health 18(4):1839

Shvetsova O, Adeel AB, Catalano M, Catalano O, Giannelli F, Muftuoglu E et al (2020) Institutional origins of COVID-19 public health protective policy response (PPI) data set v. 1.2-regional US and Canada

Tahmasebi P, Shokri-Kuehni SM, Sahimi M, Shokri N (2020) How do environmental, economic and health factors influence regional vulnerability to COVID-19? MedRxiv. https://doi.org/10.1101/2020.04.09.20059659

Uğur NG, Akbıyık A (2020) Impacts of COVID-19 on global tourism industry: a cross-regional comparison. Tourism Manage Perspect 36:100744

Wilson DH, Johnson BA, Stokan E, Overton M (2020) Institutional collective action during Covid-19: lessons in local economic development. Public Adm Rev 80 (5):862–865

17 Public Networks and Keeping the Support: Exploring a Better Governance for a Healthier Future

Governments are true reflections of their people.
—Ali Cheshmehzangi, 2014

17.1 A Brief Exploration: Public Networks for Healthy Society

Public networks include a wide range of networks from social networks to transportation networks and services, altogether resembling the complex network systems of cities and communities globally. Their governance is critical in facing natural disasters, wars, conflicts, and health crises like the COVID-19 pandemic. Both security and formalisation of public networks are essential to enhance the relationship between them, breaking the barriers and enable collaborative networks (Szabo 1997). In particular, the correlation between social networks and public action in urban governance is regarded as an essential strategy to enhance city and community livelihood (Beall 2001) and sustain urban health. Under the topic of "*the importance of being connected*", Payre (2010) argues about city networks and urban governance matters. He also explores the interurban networks that are run by public and private actors of cities and city governments (ibid). This also reflects on earlier discourse on public-sector organisational networks, such as those related to the public domain and service-delivery matters that help everyday city operations (Provan and Milward 2001). Other aspects, such as energy supplies, power, water, and traffic networks, are all relevant to the study of public networks (Ma et al. 2011). However, more importantly, we refer to their governance to be related to enhancing the city's smartness, resilience, and sustainability. Some arguments also exist regarding transnational city networks, which could help to form sustainability-oriented networks and towards organisational innovations (Keiner and Kim 2007). These studies are utilised for city administrators to create a process that involves both public and private stakeholders (ibid). The advancing urban technological networks add to these relatively smarter governance mechanisms (Kaika and Swyngedouw 2000), which have developed even more in the last two decades or so.

In more recent studies, the intersection between public network studies and smart governance is recognised as a new area of "smart city network governance" (Palomo-Navarro and Navío-Marco 2018), which also relates to shared-governance networks and governing the city's multiple public networks. The shared-governance networks aspect plays a major part in creating network partners and developing links between various stakeholders (Cristofoli et al. 2014). In smart city dashboards, we see growing demand for evaluating and optimising public

A. Cheshmehzangi, *Urban Health, Sustainability, and Peace in the Day the World Stopped*, Sustainable Development Goals Series, https://doi.org/10.1007/978-981-16-4888-5_17

networks, which help to enhance planning decision-making processes and city management (Suakanto et al. 2013). For instance, in the European National Healthy City Networks, we see growing attention to local public improvements from both structures and processes perspectives (Heritage and Green 2013). The same applies in other contexts and (for) other sectors, such as public transport networks (Lu and Shi 2007), where we see inevitable growth that requires inexorable institutional support.

In addition, the emergence of mobile social networks has played a significant part in shared information platforms (Humphreys 2010), the use of ICT-mediated platforms, and enhancing the city's capacity for data collection and usage. In this regard, the networks of public policies help to enhance local governance (Giraldo et al. 2009) and create a better approach to consider multiscalar thinking in urban sustainability studies (Cheshmehzangi et al. 2021). By leading to smart city networks, such an approach could enable to better understand visuospatial networks, socio-spatial networks, etc. (Rassia and Pardalos 2017). In smart or sustainable–smart city movements, we see the emergence of such public network consideration and integration (Anastasiadou and Vougias 2019; Frez et al. 2019). Nonetheless, smart urban governance (Meijer and Bolívar 2016), related to such research areas, has become even more important to study public networks in cities. For instance, some studies focus on various public networks in smart cities (Anthopoulos and Vakali 2012) and suggest new methods of governance networks (Anand and Navío-Marco 2018; Docherty et al. 2018; Šiugždinienė et al. 2019; Tuyls and Pera 2019; Yahia et al. 2019; Nesti and Graziano 2020; Russell 2020). These studies represent true examples of emerging public network governance studies, theories, and practices from the smart–sustainable viewpoints.

It is also debated that public networks could support urban diversity from the socio-spatial perspective (Pinto and Remesar 2012; Cheshmehzangi and Li 2020). The study by Cristofoli et al. (2014) suggests the combination of governance, management, and performance in public networks, reflecting on overarching areas of public administration and public policy (Koliba et al. 2018). Hence, this chapter focuses on key aspects of collaborative innovation in governance networks, which are discussed in correlation with public management reforms (Sørensen and Torfing 2017), shared-governance network and smart city governance (Palomo-Navarro and Navío-Marco 2018), facilitating public condensation of political decisions (Sørensen and Torfing 2005), and the governance of public policy networks (Löffler 2015). Furthermore, we zoom into the importance of public networks because of sound institutional structures. We specifically explore the significance of public networks for a healthy society and discuss more on institutional support that is needed for peacebuilding processes.

17.2 Reflecting on How Cities Reacted During the Pandemic

As evidenced throughout the COVID-19's multiple phases, public networks played a major part in city operations as well as the management of adversities caused by the pandemic. The crossover between physical domain, political domain, and public domain pushes us to reflect on urban management scenarios (Cheshmehzangi 2020a, b) and consider a better understanding of urban governance for public networks. From spatial networks (Pujari ad Shekatkar 2020) to social networks (Kuchler et al. 2020), as well as transportation networks and services networks (Cheshmehzangi 2020b), we see opportunities for enhancing the networks with shared plans and better resilience (Bai et al. 2020). Thus, as we witnessed, many cities considered adaptive measures (Cheshmehzangi 2020c) in the governance and management of their public networks. Such an approach helped enhance performances or verify deficiencies, shortfalls, and shortages that were neglected prior to the pandemic.

Globally, the COVID-19 pandemic had a significant impact on the physical domain and public networks of cities. For instance, the continuing distrust of public transportation led to further complications in urban systems and networks (Cheshmehzangi 2020d). One of the main implications is energy use (ibid), as the decline in public transportation use indicates an increase in private car use. Such impact on public transportation use means an irrepressible decline in the financial and institutional support of such public networks. As evident during the pandemic, we notice weakening institutional support for the sustainability of public services and networks. This weakening could cost countries/cities more than just a temporary decline of their institutional structures. Furthermore, we see larger impacts on major global movements, such as shared facilities and shared public networks, which may require more time for further readjustments.

An example of a positive response from cities on handling the COVID-19 immediate effects was some of the success stories regarding creating healthcare and hospital networks (Rubin et al. 2020). Such examples used shared data platforms, ICT, and other new methods in enhancing the essential services. This global crisis also opened up opportunities for smart-resilient network systems, such as those for long studied in the smart city networks, information-based methods, etc. (Allam and Jones 2020). The use of smart technologies to enhance city networks (Inn 2020) was also a common example in developed and rapidly developing contexts. For instance, artificial intelligence (AI) for urban health monitoring and management (ibid) is just one of the few transformative incidents from the COVID-19 pandemic. Many new ideas or initiatives also review the COVID-19 pandemic with the aim to enhance city networks, develop new smart city concepts, use digital technology, and improve the city's resilience (Costa and Peixoto 2020; Rahman et al. 2020; Su et al. 2021). The push about the use of surveillance technology (Sonn and Lee 2020), in the name of safety and security (Cheshmehzangi 2020a), has already made significant changes to public network operations and management. We see a growing demand for surveillance and monitory systems at the city and intra-city public networks, which suggest a shift in data collection and usage in the near future.

17.3 Suggestions Against the Common Flaws

To respond more reflectively to the COVID-19 adversities, we ought to better understand public health networks and other essential networks that seem ever important in everyday operations (Cheshmehzangi 2020a). Proposals for network interventions mean opportunities for network solutions (Young et al. 2021) and enabling us to be trained and better prepared for future events (Al Nsour et al. 2020). These approaches could also help us better prepare through urban resilience (Cheshmehzangi 2020c), which could happen by optimising our public network operations, institutional structures, and support that may be needed in achieving these.

In addition, we expect to see further development of surveillance-assisted public networks and services, such as public health services (Kim et al. 2021) and public transportation services (Tirachini and Cats 2020). Shared data platforms are expected to grow more in the future, which may be beneficial to inform specific action plans and help us to develop cross-sectoral systems. We are likely to evolve more with a smarter use of public databases and "*generate a deduced network*" (Chookajorn 2020), which could be used for future collaborative networks of various sorts (Liu et al. 2020) and potential emergency action plans. Nonetheless, we also see a growing concern about cybersecurity matters (Wirth 2020), which are likely to remain as the nemesis of public databases and issues associated with data usage and management. Thus, urban health strategies are expected to have robust action plans, ensuring not only the safety and security of operations but also data safety and their optimum use. We look into this with a dubious viewpoint as we witnessed many cyberattacks and data misuse, and mismanagement during the COVID-19 pandemic.

In addition, growing attention is expected to be on the governance of public networks, including their daily operations, their risk management plans, their priority plans, etc. (Cheshmehzangi 2020a). Our public administrations should focus on better governance of public networks, allowing for fairer situations between private and public stakeholders. Such an approach could lead to better sustainment of city networks, their operations, and the institutional support that they require for their healthy continuity. In the future, we need more interventions and innovations that just technology-based and information-based approaches. We see more inclusivity in public interventions than those technology-based approaches led by governments. For instance, our urban social networks need to get refreshed through institutional restructuring processes. In doing so, we could evolve out of conventional provisions of public services and innovate with better essential services and public networks. By achieving this, we may be able to simply break the long-term barriers and silos between different departments and units of the city governments. The implications would then be on common practices, which would possibly advance our public networks through less corrupted institutional support. This could create better emergency plans for public networks and enable us to have healthier financial support for specific sectors. By doing this, we could also enhance fairness in public–private partnerships, which could create healthier competition than implementing top-down governance measures.

Lastly, we reflect on common financial challenges, which often weaken public networks and services. Our first reflective viewpoint on this matter would be on low-level corruption, which has been thoroughly discussed in Chap. 20. As a crossover between economy and institutional structures, we note the importance of financial and institutional support for public networks, without which their operations will never be healthy, and nor will they ever be sustainable. Thus, in order to achieve an effective health–sustainability–peace (HSP) nexus, cities and communities ought to restructure their institutional structures if needed, develop their institutional frameworks if not existent, and consider collaborative innovation in governance networks if not yet decided.

17.4 A Summary

This chapter serves as a platform where we explored the overarching topic of public networks from the institutional perspective. We highlighted the critical aspect of governance, not just because it is weakening but also because some of our thinking remains conventional, unviable, and unfair. If we opt to move through inclusiveness, then we need to go through significant institutional restructuring processes, with the central purpose of reducing corruption levels, enhancing partnerships, and optimising fairness and justice in our public networks. For a robust HSP nexus in cities and communities, we ought to value our public networks but also allow for innovative interventions and not just towards technology-based and information-based platforms.

As we close our investigation of various aspects under the institutional sustainability dimension, we highlight many crossovers between this dimension and the other dimensions of economic, social, and environmental. We see growing demand for governance interventions and suggest new perspectives on technical and technological pathways. Thus, in the next part of the book, we delve into the technical dimension, which would highlight some of the discussions that have already been initiated in this chapter. To follow up, we raise our concerns on our over-reliance on technologies and technology-based platforms and suggest pathways that could be beneficial for future smart-resilience directions. The institutional dimension, covered in this part of the book, offers us many overlapping ideas, which indicate that the HSP nexus is not a myth but is a possibility through healthy and fair governance. However, we note—with a moral consciousness—that our current political and collective willpower and values are far from this ideal vision.

References

Al Nsour M, Bashier H, Al Serouri A, Malik E, Khader Y, Saeed K et al (2020) The role of the global health development/eastern mediterranean public health network and the eastern mediterranean field epidemiology training programs in preparedness for COVID-19. JMIR Public Health Surveill 6(1):e18503

Allam Z, Jones DS (2020, March) On the coronavirus (COVID-19) outbreak and the smart city network: universal data sharing standards coupled with artificial intelligence (AI) to benefit urban health monitoring and management. In: Healthcare, vol 8, No 1, p 46. Multidisciplinary Digital Publishing Institute

Anand PB, Navío-Marco J (2018) Governance and economics of smart cities: opportunities and challenges. Telecommun Policy 42(10):795–799

Anastasiadou K, Vougias S (2019) "Smart" or "sustainably smart" urban road networks? The most important commercial street in Thessaloniki as a case study. Transp Policy 82:18–25

Anthopoulos LG, Vakali A (2012, May) Urban planning and smart cities: Interrelations and reciprocities. In: The future internet assembly. Springer, Berlin, Heidelberg, pp 178–189

Bai X, Nagendra H, Shi P, Liu H (2020) Cities: build networks and share plans to emerge stronger from COVID-19

Beall J (2001) From social networks to public action in urban governance: where does benefit accrue? J Int Dev 13(7):1015–1021

Cheshmehzangi A (2020a) The city in need: urban resilience and city management during disruptive disease outbreak events. Springer, Singapore

Cheshmehzangi A (2020b) Reflection on disruptions: managing the city in need, saving the city in need. In: Cheshmehzangi A (ed) The city in need: urban resilience and city management during disruptive disease outbreak events, Springer, Singapore, pp 137–283

Cheshmehzangi A (2020c) Preparedness through urban resilience. In: Cheshmehzangi A (ed) The city in need: urban resilience and city management during disruptive disease outbreak events. Springer, Singapore, pp 41–104

Cheshmehzangi A (2020d) COVID-19 and household energy implications: what are the main impacts on energy use? Heliyon 6(10):e05202

Cheshmehzangi A, Li HMA (2020) Innovation through urban diversity and achieving comprehensive sustainable urbanism from a community-oriented approach. Curr Urban Stud 8(02):222

Cheshmehzangi A, Flynn A, Tan-Mullins M, Xie L, Deng W, Mangi E, Chen W (2021) From eco-urbanism to eco-fusion: an augmented multi-scalar framework in sustainable urbanism. Sustainability 13 (4):2373

Chookajorn T (2020) Evolving COVID-19 conundrum and its impact. Proc Natl Acad Sci USA 117(23):12520

Costa DG, Peixoto JPJ (2020) COVID-19 pandemic: a review of smart cities initiatives to face new outbreaks. IET Smart Cities 2(2):64–73

Cristofoli D, Markovic J, Meneguzzo M (2014) Governance, management and performance in public networks: How to be successful in shared-governance networks. J Manage Governance 18(1):77–93

Docherty I, Marsden G, Anable J (2018) The governance of smart mobility. Transp Res Part A: Policy Pract 115:114–125

Frez J, Baloian N, Pino JA, Zurita G, Basso F (2019) Planning of urban public transportation networks in a smart city. J UCS 25(8):946–966

Giraldo G, Morales J, Granada Vahos J (2009). Las Políticas Públicas Territoriales Como Redes De Política Pública Y Gobernanza Local: La Experiencia De Diseño Y Formulación De Las Políticas Públicas Sobre Desplazamiento Forzado En El Departamento De Antioquia Y La Ciudad De Medellín (Territorial public politics as networks of public policies and local governance: the experience of the design and formulation of the public policies on forced displacement in the department of Antioquia and the city of Medellín). Estudios Políticos ISSN:0121-5167

Heritage Z, Green G (2013) European national healthy city networks: the impact of an elite epistemic community. J Urban Health 90(1):154–166

Humphreys L (2010) Mobile social networks and urban public space. New Media Soc 12(5):763–778

Inn TL (2020) Smart city technologies take on COVID-19. World Health 841. Available from: https://www.ncbi.nlm.nih.gov/pmc/articles/PMC7188041/https://penanginstitute.org/wp-content/uploads/2020/03/27_03_2020_TLI_download.pdf

Kaika M, Swyngedouw E (2000) Fetishizing the modern city: the phantasmagoria of urban technological networks. Int J Urban Reg Res 24(1):120–138

Keiner M, Kim A (2007) Transnational city networks for sustainability. Eur Plan Stud 15(10):1369–1395

Kim L, Garg S, O'Halloran A, Whitaker M, Pham H, Anderson EJ et al (2021) Risk factors for intensive care unit admission and in-hospital mortality among hospitalized adults identified through the US coronavirus disease 2019 (COVID-19)-associated hospitalization surveillance network (COVID-NET). Clin Infect Dis 72(9):e206–e214

Koliba CJ, Meek JW, Zia A, Mills RW (2018) Governance networks in public administration and public policy. Routledge, New York and London

Kuchler T, Russel D, Stroebel J (2020) The geographic spread of COVID-19 correlates with the structure of social networks as measured by Facebook (No. w26990). National Bureau of Economic Research

Liu J, Hao J, Shi Z, Bao HX (2020) Building the COVID-19 collaborative emergency network: a case study of COVID-19 outbreak in Hubei Province, China. Nat Hazards 104(3):2687–2717

Löffler E (2015) Public governance in a network society. In: Public management and governance. Routledge, New York and London, pp 233–248

Lu H, Shi Y (2007) Complexity of public transport networks. Tsinghua Sci Technol 12(2):204–213
Ma K, Wang Z, Jiang J, Zhu G, Li W (2011) Power law and small world properties in a comparison of traffic city networks. Chin Sci Bull 56(34):3731–3735
Meijer A, Bolívar MPR (2016) Governing the smart city: a review of the literature on smart urban governance. Int Rev Admin Sci 82(2):392–408
Nesti G, Graziano PR (2020) The democratic anchorage of governance networks in smart cities: an empirical assessment. Public Manag Rev 22(5):648–667
Palomo-Navarro A, Navío-Marco J (2018) Smart city networks' governance: the Spanish smart city network case study. Telecommun Policy 42(10):872–880
Payre R (2010) The importance of being connected. City networks and urban government: Lyon and Eurocities (1990–2005). Int J Urban Reg Res 34(2):260–280
Pinto AJ, Remesar A (2012) Public space networks as a support for urban diversity. Open House Int 37(2). https://doi.org/10.1108/OHI-02-2012-B0003/full/html#loginreload
Provan KG, Milward HB (2001) Do networks really work? A framework for evaluating public-sector organizational networks. Public Adm Rev 61 (4):414–423
Pujari BS, Shekatkar SM (2020) Multi-city modeling of epidemics using spatial networks: Application to 2019-nCov (COVID-19) coronavirus in India. *Medrxiv*. https://doi.org/10.1101/2020.03.13.20035386v1
Rahman MM, Manik MMH, Islam MM, Mahmud S, Kim JH (2020, September) An automated system to limit COVID-19 using facial mask detection in smart city network. In: 2020 IEEE international IOT, electronics and mechatronics conference (IEMTRONICS). IEEE, New York, pp 1–5
Rassia STh, Pardalos PM (eds) (2017) Smart city networks: through the internet of things. Part of the Springer optimisation and its applications book series (SOIA, volume 125), Springer International Publishing
Rubin GA, Wan EY, Saluja D, Thomas G, Slotwiner DJ, Goldbarg S et al (2020) Restructuring electrophysiology during the COVID-19 pandemic: a practical guide from a New York City hospital network, Critical pathways in cardiology. Available from: https://www.ncbi.nlm.nih.gov/pmc/articles/PMC7188041/
Russell H (2020) Sustainable urban governance networks: data-driven planning technologies and smart city software systems. Geopol History Int Relat 12(2):9–15
Šiugždinienė J, Gaulė E, Rauleckas R (2019) In search of smart public governance: the case of Lithuania. Int Rev Adm Sci 85(3):587–606
Sonn JW, Lee JK (2020) The smart city as time-space cartographer in COVID-19 control: the South Korean strategy and democratic control of surveillance technology. Eurasian Geogr Econ 61(4–5):482–492
Sørensen E, Torfing J (2017) Metagoverning collaborative innovation in governance networks. Am Rev Public Admin 47(7):826–839
Sørensen E, Torfing J (2005) The democratic anchorage of governance networks. Scand Polit Stud 28(3):195–218
Su Z, McDonnell D, Bentley BL, He J, Shi F, Cheshmehzangi A et al (2021) Addressing Biodisaster X Threats with AI and 6G Technologies. J Med Int Res (in press)
Suakanto S, Supangkat SH, Saragih R (2013, June) Smart city dashboard for integrating various data of sensor networks. In: International conference on ICT for smart society. IEEE, New York, pp 1–5
Szabo N (1997) Formalizing and securing relationships on public networks. First Monday. Available from: https://firstmonday.org/ojs/index.php/fm/article/view/548
Tirachini A, Cats O (2020) COVID-19 and public transportation: current assessment, prospects, and research needs. J Public Transp 22(1):1
Tuyls R, Pera A (2019) Innovative data-driven smart urban ecosystems: environmental sustainability, governance networks, and the cognitive Internet of Things. Geopol History Int Relat 11(1):116–121
Wirth A (2020) Cyberinsights: COVID-19 and what it means for cybersecurity. Biomed Instrum Technol 54 (3):216–219
Yahia NB, Eljaoued W, Saoud NBB, Colomo-Palacios R (2019) Towards sustainable collaborative networks for smart cities co-governance. Int J Inform Manage 102037
Young LE, Sidnam-Mauch E, Twyman M, Wang L, Xu JJ, Sargent M et al (2021) Disrupting the COVID-19 misinfodemic with network interventions: network solutions for network problems. Am J Public Health 111(3):514–519

Part VI

Sustainability and Peace: "Urban Health from the Technical Dimension"

In the sixth and final part of studying sustainability and peace, we present three primary aspects representing urban health from the technical dimension.

The first aspect is technology availability and accessibility, primarily discussed in the face of the growing digital divide in our contemporary society. These are discussed in Chap. 18, focusing on addressing digital divide issues while achieving urban health from the technical and technological perspectives. The second aspect is focused on asset and facilities management from a technical standpoint. In this aspect, we highlight the importance of the two in maintaining a good health–sustainability–peace nexus in cities and communities. These are discussed in Chap. 19, overtly exploring technical viewpoints and related factors that may be suitable for the better management and containment of future pandemics. The third aspect is related to smart platforms and technical solutions, another crossover between technical and technological directions. We explicitly explore the topic of smart resilience and suggest ways in which it could be achieved in practice. This critical aspect is presented in Chap. 20, which is the closing chapter of all six dimensions before making the conclusions.

18 Technology Availability and Accessibility: Dealing with the Issues of Digital Divide

Our nature dictates the continuity of barriers between classes, genders, races, etc.
—Ali Cheshmehzangi, 2015

18.1 Knowing More About Technology Availability and Accessibility

For long, the relationships between multiple innovations of technology, technical, and organisational have led to economic and societal growth directions (Sanidas 2004), which also indicates a better technical intelligence in the performance of our working environments (Ashton ad Stacey 1995). Various technical models of technologies help us to verify various aspects of technology security, technical foundations, and their integration in everyday operations (Stoneburner 2001). As part of the technological advancement and technical change in cities, we see a tangible shift towards technical entrepreneurship and implications for management (Oakey 2003). Therefore, the integration of technologies in technical processes suggests various new strategies that enable better use of data, open up room for innovation, and improve communications, competency, and competitiveness between various stakeholders. In particular, we stress technical communication and information technology (Hart-Davidson 2001), which for a long time has enabled us to boost technology diffusion (Soete and Turner 1984), technology adoption (Nambisan and Wang 1999), and intervene with innovation.

Associated with the technical sustainability perspective, we see a growing demand also in technical peace research (Altmann 2019). This fact indicates a further development of the area beyond just military technologies (Sherwin 1956; Reuter et al. 2020) and towards information technology for peace and security (Reuter 2019), ICT for peace (Stauffacher et al. 2005; Reuter 2020), and utilising technology for peacekeeping strategies (Convergne and Snyder 2015) and peace operations (Dorn 2011). In addition, technology has been embedded as part of sustainability assessment indicators (Dunmade 2002), which has helped directions in the development of socio-technological and socio-technical pathways, some that encourage innovation in enhancing sustainability measures (Brochner et al. 1999), and some that have created new paradigms of development. There are also challenges and concerns about technological regime shifts (Kemp 1994), which signify the importance of technology–technical support (Christiansen 1997) in achieving sustainability transitions.

A. Cheshmehzangi, *Urban Health, Sustainability, and Peace in the Day the World Stopped*, Sustainable Development Goals Series, https://doi.org/10.1007/978-981-16-4888-5_18

From more popular socio-technical transitions for sustainability, we see the role of technology in "*achieving transitions towards sustainable development*" in various areas, such as in the area of energy (Langhelle et al. 2019). The use of integrated sustainability frameworks has been helpful for us to invest in and apply various technologies in everyday practices (Dao et al. 2011). The approach has led to improvements in technological practices (Davison 2001) and achieving sustainability in most cases. However, it also leads us to govern technology for sustainability (Murphy 2012), particularly from the technical perspective. The use and application of technologies alone are not suitable to promote health–sustainability–peace directions; hence, we ought to realise how we could govern them and allow for better availability and accessibility. In this regard, we see a growing demand for blindfolded use of technologies and replacing the conventional methods, but without considering how they should be integrated, assessed, and (ultimately) governed. In order to accelerate socio-technical transitions for sustainability (Langhelle et al. 2019), we cannot simply rely on the deployment of technologies to city operations. Instead, we should have a better understanding of our socio-technical systems and nature of technologies and processes and comprehensively consider system innovation for better resilience and sustainability governance. For the latter, we need to think of more than just technology-focused practices (Smith and Stirling 2008) and evaluate opportunities for hybrid models, practices, and platforms.

From the technical perspective, this chapter focuses on technology availability and accessibility, which we believe are essential to promote peace and prosperity. Particularly, we debate the growing digital divide issues and those that could become barriers than enablers for sustainability. In the studies of the transnational city for networks for sustainability, we see a broader understanding of technical subjects and attributes, which suggests the importance of the technical dimension in understanding sustainability transitions (Keiner and Kim 2007). Even the pathways for environmental sustainability would entail technical factors regarding peace and prosperity improvements (World Bank 2019). Thus, for a better sustainability management approach, we out to combat "*key technical, financial, organisational, and political challenges*" (Cohen 2011). It is only in this process that we could appreciate the technicalities over technologies and ensure green entrepreneurship is based on technology integration and not just technologies. As Amadei (2021) highlights in the study of the sustainability–peace nexus, we could see the importance of both technical and non-technical processes. But, we also add that in dealing with sustainability trade-offs, we cannot neglect the role of technical sustainability aspects, including those that are essential to enhance partnerships, equalities, and prosperity.

As highlighted throughout the book, all sustainability dimensions are equally important for the health–sustainability–peace (HSP) nexus in cities and communities. However, when it comes to the technical dimension, we note the role of technologies or technical–technology nexus and those relevant to sustainability directions. In this regard, we first look into technology availability and accessibility, ensuring that educational and social factors are also included in reducing the digital divide in society (Li and Ranieri 2013). Arguments on technical appropriations of technical capital (Brock et al. 2010) also suggest the role of ICT and activating cultural and technical capital in enhancing cities and communities. However, to redress the digital divide issues, we see more crossovers emerging between the social and technical dimensions in achieving a fair HSP nexus in cities. These are commonly discussed through the analysis of information systems (Kvasny and Keil 2006), evaluating technical issues (Light 2001) and technical skills (Norris 2020), and matters of social inclusion (Warschauer 2012). Therefore, this chapter aims to address some of the common—and growing—issues of our contemporary society. By exploring the ongoing progression from the digital divide to digital inequality (DiMaggio and Hargittai 2001), we need to understand technologies, communities, and public policies (Servon 2008). Thus, our goal is to reflect on digital divide issues while achieving urban health from the technical and technological perspectives. In doing so, we aim to respond to the current situation of the pandemic and what we expect to happen when it is over. The impacts, as we witness to date, will be widespread, and our approach to peacemaking should happen now and not later.

18.2 Reflecting on How Cities Reacted During the Pandemic

The COVID-19 pandemic became a good platform to experiment with innovative technological solutions to manage the situation. A common example is the use of information technology, which was used widely in multiple sectors and mainly the healthcare units in cities (Cheshmehzangi 2020a; Salway et al. 2020). Thus, we see an opportunity to develop new technologies, speed up their integrations, and manage scenarios (Cheshmehzangi 2020b, c). In this process, we have detected a sort of transformation (Cheshmehzangi 2021), which was already in the process through gradual transitions. While there are some technological progressions during the pandemic, many scholars also note technological challenges (Greenspan et al. 2020) and suggest a better assessment of such innovations (Gyimah 2020). In this regard, technological advancements may differ from context to context and sector to sector. In particular, the use of technologies in developing practical considerations and community resilience plans (Conroy et al. 2020) is crucial to technical evaluation factors. As shown in the case of the Indian response during the first phase of the pandemic, technological interventions were combined with policy development, which helped to catalyse the recovery process at first (Goel et al. 2021). Nevertheless, the country's later situation indicated that technological innovation alone could not be simply sustained in a longer process when needed. Our concern is mainly on technological self-governance mechanisms (Shin 2021), which ought to be understood before any further implementations.

In many cases, there are issues of accessibility and affordability to certain technologies (even if small) that create a larger digital divide in society (Reddick et al. 2020). For example, many researchers and officials have widely welcomed the use of the health QR code system in China. However, it also affected certain social groups, such as the elderly and those with no access to the internet or smart phones (Wang and Jia 2021). Some studies also indicate growing issues of cyberculture in places where the digital divide has been developing for a long (Naseer and Aktaş 2019a, b). In such studies, digital literacy is considered a matter of concern in dealing with pandemics such as COVID-10 (ibid). Yet, the widening gap between old and young populations could lead to further division in society, which means a more expansive digital divide (Yao et al. 2021), which may become a challenge after the pandemic. Issues of social exclusion and health disparities are yet to add to this rally (Gibson et al. 2020), which are likely to increase digital exclusion (Yao et al. 2021) or become a shift towards digitalisation and digitisation (Cheshmehzangi 2021). In some contexts, the forced digital applications become the common norm, and this may not truly respond to issues of accessibility and affordability. It may in some ways respond to the availability of technologies, but the issues may not necessarily help social inclusion directions and those that are needed for inclusive and sustainable communities.

In addition, mobile health technology was also among those technological innovations that helped boost smart health platforms, using Internet+ methods and overcoming traditional data collection approaches (Wu et al. 2021). The same applies to contact tracing work, using QR code provided by the government for the use of track and recording processes in countries like New Zealand (Sholeh et al. 2020). Such digitalisation and automation approaches could be helpful, if not at the cost of widening the digital skills gaps (Kapetaniou 2020). They could only be as impactful when it comes to specific sectors. For instance, an adverse impact was experienced regarding online education approaches, which widened the digital divide in poorer communities

(Mishra 2020; Azubuike et al. 2021; Ware 2021). Such inequalities severely impact social power and inclusion movements, which again should be understood as more than just access to the Internet (Lai and Widmar 2021). In fact, the impact of the digital divide on people could be seen even in developed countries and in cities of higher-level prosperity (Ramsetty and Adams 2020; Vogels et al. 2020; Watts 2020; Aissaoui 2021).

In facing the exacerbation of disparities (Bakhtiar et al. 2020), we see new digital divide issues also emerging in areas such as telework-ability (Sostero et al. 2020), digital health access (Eruchalu et al. 2021), telemedicine (Eberly et al. 2020), and education (Hussain 2020). The rural digital divide is expected to be much larger than those experienced in urban regions (Esteban-Navarro et al. 2020). We also note growing issues that could lead to new lifestyles, which may not experience a set of divergence. Such examples may include increasing demand for indoor entertainment for the future (Cheshmehzangi cr), where it would still exclude the more impoverished communities and leading to widening the digital divide in household equipment use. We also note the digital exclusion of older adults, as suggested by Seifert (2020), may become more sensitive in regard to double exclusions. This means the physical isolation and digital divide (ibid) would have longer-term effects on those vulnerable groups. Thus, there is an urgent need to evaluate the situation more cautiously than letting the current needy trends become societal needs in future.

18.3 Addressing Issues of Digital Divide

To summarise, our technical concern is mainly on availability and accessibility factors. The former is primarily related to digital literacy, affordability, and matters of choice, which indicate broader concerns in democratic societies. The latter is another matter that could mean increasing inequalities, which would put pressure on pro-poor development plans and other campaigns against poverty and social exclusion issues. As the COVID-19 pandemic continues to magnify the digital divide (Majeed et al. 2020), we see disparities emerging also in between sectors and urban systems too. The many obstacles in the use of digital technological solutions in many sectors (Giansanti et al. 2021), should be verified regarding technical issues that include accessibility and availability factors. From the digital divide in education (Rodicio-García et al. 2020) to growing digital inequalities (Zheng et al. 2021), it is necessary for us to consider other factors, such as digital democracy, digital access (de los Santos and Rosser 2020), and technical consideration of digital divide. In doing so, we may reflect on some of the growing issues with an added care and towards inclusivity and bridging between different sectors, social groups, genders, ages, etc.

Once again, we highlight trendy opinions that indicate technological advancement will become our saviour, without considering its governance, availability, accessibility, safety, security, and of course, its integration and adaptation. The COVID-19 pandemic suggests various issues related to digital divide progress, including the lockdown effects on education disengagement of the poor (Seymour et al. 2020), socio-economic injustices (Oldekop et al. 2020), special stigma, and inequalities (Aziz et al. 2020), and information technology use (Jackson et al. 2008). Nowadays, people are unwillingly signing up to many digital platforms that are exposing them to wider networks of digital fraud and cyberattacks. We see growing inequalities developing from digital divide progression, which may put us one or two generations back in regard to social inclusion considerations. Thus, it is now that we have to act to ensure an excellent technical–technological consideration is maintained, information management is secured, and adequate support is provided to those most vulnerable in our society.

18.4 A Summary

It is often challenging to summarise something as broad as what has been discussed in this chapter. We have already raised several concerns related to technical–technological issues that may widen the gap between contexts, sectors, social groups, etc. We notice a growing force behind making digital compulsory, with an easier societal acceptance. Those movements have already begun long ago, when we knowingly or unknowingly opt to share our data/information, be active in social media, have a good share of data exchanges, etc. We see these to become trends that are almost inevitable but are not necessarily healthy and sustainable for our society. The digital divide issues could lead to controversial topics related to longer-term and structured social exclusions, growing double standards or attitudes that could divide citizens into certain groups and categories, profiling individuals and specific ethnicities, and so on. These issues have for long being hidden and existent, and the current situation just enables us to see them more clearly.

As the starting aspect of the technical dimension, this chapter highlights some of the common norms and some emerging issues that may be excelled after the pandemic. In order to bridge the digital divide, we ought to understand the technological–technical nexus of progressive digitalisation, which would include its better integration, governance, and inclusion. The impacts of the COVID-19 on specific sectors should help us evaluate their sustainability and future directions. In doing so, we may be able to address information and digital literacy issues and reflect on what causes the digital divide in the first place. We could also develop more collective and inclusive approaches that could ultimately address issues of the digital divide, create all-inclusive action plans, make tangible and transformative progress in reducing the inequalities. If we do so, we could mitigate the issues of the digital divide, but we probably could not eradicate the digital divide itself.

References

Aissaoui N (2021) The digital divide: a literature review and some directions for future research in light of COVID-19. Global Knowl Memory Commun. https://doi.org/10.1108/GKMC-06-2020-0075/full/html

Altmann J (2019) Natural-science/technical peace research. In: Information technology for peace and security. Springer Vieweg, Wiesbaden, pp 39–60

Amadei B (2021) A systems approach to the sustainability–peace nexus. Sustain Sci, 1–14

Ashton WB, Stacey GS (1995) Technical intelligence in business: understanding technology threats and opportunities. Int J Technol Manage 10(1):79–104

Aziz A, Islam MM, Zakaria M (2020) COVID-19 exposes digital divide, social stigma, and information crisis in Bangladesh. Media Asia 47(3–4):144–151

Azubuike OB, Adegboye O, Quadri H (2021) Who gets to learn in a pandemic? Exploring the digital divide in remote learning during the COVID-19 pandemic in Nigeria. Int J Educ Res Open 2:100022

Bakhtiar M, Elbuluk N, Lipoff JB (2020) The digital divide: how Covid-19's telemedicine expansion could exacerbate disparities. J Am Acad Dermatol 83(5): e345–e346. Available from: https://www.ncbi.nlm.nih.gov/pmc/articles/PMC7365110/

Brochner J, Ang GK, Fredriksson G (1999) Sustainability and the performance concept: encouraging innovative environmental technology in construction. Build Res Inform 27(6):367–372

Brock A, Kvasny L, Hales K (2010) Cultural appropriations of technical capital: black women, weblogs, and the digital divide. Inf Commun Soc 13(7):1040–1059

Cheshmehzangi A (2020) The city in need: urban resilience and city management during disruptive disease outbreak events. Springer, Singapore

Cheshmehzangi A (2021) From transitions to transformation: a brief review of the potential impacts of COVID-19 on boosting digitization, digitalization, and systems thinking in the built environment. J Build Construct Plan Res 9(01):26

Cheshmehzangi A (2020b) Reflection on disruptions: managing the city in need, saving the city in need. In: Cheshmehzangi A (ed) The city in need: urban resilience and city management during disruptive disease outbreak events. Springer, Singapore, pp 137–283

Cheshmehzangi A (2020c) Preparedness through urban resilience. In: Cheshmehzangi A (ed) The city in need: urban resilience and city management during disruptive disease outbreak events. Springer, Singapore, pp 41–104

Cheshmehzangi A (2020d) COVID-19 and household energy implications: what are the main impacts on energy use? Heliyon 6(10):e05202

Christiansen E (1997) Gardening: a metaphor for sustainability in information technology-technical support.

In: An ethical global information society. Springer, Boston, MA, pp 171–185
Cohen S (2011) Sustainability management: lessons from and for New York City, America, and the planet. Columbia University Press
Conroy KM, Krishnan S, Mittelstaedt S, Patel SS (2020) Technological advancements to address elderly loneliness: practical considerations and community resilience implications for COVID-19 pandemic. Work Older People 24(4). https://doi.org/10.1108/WWOP-07-2020-0036/full/html
Convergne E, Snyder MR (2015) Making maps to make peace: geospatial technology as a tool for UN peacekeeping. Int Peacekeep 22(5):565–586
Dao V, Langella I, Carbo J (2011) From green to sustainability: information technology and an integrated sustainability framework. J Strateg Inf Syst 20 (1):63–79
Davison A (2001) Technology and the contested meanings of sustainability. Suny Press, Albany, New York
de los Santos GE, Rosser W (2020) COVID-19 shines a spotlight on the digital divide. Change: Maga High Learn 53(1):22–25
DiMaggio P, Hargittai E (2001) From the 'digital divide' to 'digital inequality': studying Internet use as penetration increases. Princeton: Center Arts Cultural Policy Stud Woodrow Wilson School, Princeton Univ 4(1):4–2
Dorn AW (2011) Keeping watch: monitoring, technology and innovation in UN peace operations. United Nations University Press, Tokyo, p 209
Dunmade I (2002) Indicators of sustainability: assessing the suitability of a foreign technology for a developing economy. Technol Soc 24(4):461–471
Eberly LA, Khatana SAM, Nathan AS, Snider C, Julien HM, Deleener ME, Adusumalli S (2020) Telemedicine outpatient cardiovascular care during the COVID-19 pandemic: bridging or opening the digital divide? Circulation 142(5):510–512
Eruchalu CN, Pichardo MS, Bharadwaj M, Rodriguez CB, Rodriguez JA, Bergmark RW et al (2021) The expanding digital divide: digital health access inequities during the COVID-19 pandemic in New York City. J Urban Health 98(2):183–186
Esteban-Navarro MÁ, García-Madurga MÁ, Morte-Nadal T, Nogales-Bocio AI (2020, December) The rural digital divide in the face of the COVID-19 pandemic in Europe—recommendations from a scoping review. In: Informatics, vol 7, No 4, p 54. Multidisciplinary Digital Publishing Institute
Giansanti D, Velcro G (2021, April) The digital divide in the era of COVID-19: an investigation into an important obstacle to the access to the mHealth by the citizen. In: Healthcare, vol 9, No 4, p 371. Multidisciplinary Digital Publishing Institute
Gibson A, Bardach SH, Pope ND (2020) COVID-19 and the digital divide: will social workers help bridge the gap? J Gerontolog Soc Work, 1–3
Goel I, Sharma S, Kashiramka S (2021) Effects of the COVID-19 pandemic in India: an analysis of policy and technological interventions. Health Policy Technol 10(1):151–164
Greenspan H, Estépar RSJ, Niessen WJ, Siegel E, Nielsen M (2020) Position paper on COVID-19 imaging and AI: From the clinical needs and technological challenges to initial AI solutions at the lab and national level towards a new era for AI in healthcare. Med Image Anal 66:101800
Gyimah N (2020) Assessing technological innovation on education in the world of coronavirus (COVID-19). Available at SSRN 3670389, from online source: https://papers.ssrn.com/sol3/papers.cfm?abstract_id=3670389
Hart-Davidson W (2001) On writing, technical communication, and information technology: the core competencies of technical communication. Tech Commun 48(2):145–155
Hussain TA (2020) Education and COVID-19 in Nigeria: tackling the digital divide. SOAS Blog, Available from: https://www.soas.ac.uk/blogs/study/covid-19-nigeria-digital-divide/
Jackson LA, Zhao Y, Kolenic A III, Fitzgerald HE, Harold R, Von Eye A (2008) Race, gender, and information technology use: the new digital divide. Cyberpsychol Behav 11(4):437–442
Kapetaniou C (2020) Learning in a pandemic: closing the digital skills gap during COVID-19. Published by Nesta, London, 34 pages, retrieved from VOCED plus. Available from: https://www.nesta.org.uk/report/learning-pandemic-closing-digital-skills-gap-during-covid-19/
Keiner M, Kim A (2007) Transnational city networks for sustainability. Eur Plan Stud 15(10):1369–1395
Kemp R (1994) Technology and the transition to environmental sustainability: the problem of technological regime shifts. Futures 26(10):1023–1046
Kvasny L, Keil M (2006) The challenges of redressing the digital divide: a tale of two US cities. Inf Syst J 16 (1):23–53
Lai J, Widmar NO (2021) Revisiting the digital divide in the COVID-19 era. Appl Econ Perspect Policy 43 (1):458–464
Langhelle O, Meadowcroft J, Rosenbloom D (2019) Politics and technology: deploying the state to accelerate socio-technical transitions for sustainability. In: What next for sustainable development? Edward Elgar Publishing, Cheltenham
Li Y, Ranieri M (2013) Educational and social correlates of the digital divide for rural and urban children: a study on primary school students in a provincial city of China. Comput Educ 60(1):197–209

Light J (2001) Rethinking the digital divide. Harv Educ Rev 71(4):709–734

Majeed A, Msile E, Coronini-Cronberg S (2020) Covid-19 is magnifying the digital divide. BMJ Opin. Availale from: https://laptrinhx.com/news/covid-19-is-magnifying-the-digital-divide-pGZbpeA/amp/

Murphy J (2012) Governing technology for sustainability. Routledge, New York and London

Nambisan S, Wang YM (1999) Technical opinion: roadblocks to Web technology adoption? Commun ACM 42(1):98–101

Naseer M, Aktaş C (2019a) Bridging the gap with QR codes: QR codes for enhancing cyberculture in Istanbul. In: Handbook of research on social inequality and education. IGI Global, pp 366–382

Naseer M, Aktaş C (2019b) Bridging gap With QR codes: digital literacy among youth in Istanbul. In: Handbook of research on media literacy research and applications across disciplines. IGI Global, pp 204–219

Norris P (2020) The digital divide. Routledge, New York and London, pp 273–286

Oakey RP (2003) Technical entreprenenurship in high technology small firms: some observations on the implications for management. Technovation 23 (8):679–688

Oldekop JA, Horner R, Hulme D, Adhikari R, Agarwal B, Alford M et al (2020) COVID-19 and the case for global development. World Dev 134:105044

Ramsetty A, Adams C (2020) Impact of the digital divide in the age of COVID-19. J Am Med Inform Assoc 27 (7):1147–1148

Reddick CG, Enriquez R, Harris RJ, Sharma B (2020) Determinants of broadband access and affordability: an analysis of a community survey on the digital divide. Cities 106:102904

Reuter C (2020) Towards IT peace research: challenges at the intersection of peace and conflict research and computer science. S&F Sicherheit Und Frieden 38 (1):10–16

Reuter C, Altmann J, Göttsche M, Himmel M (2020) Natural science and technical peace research: definition, history, and current work. S&F Sicherheit und Frieden 38(1):1–4

Reuter C (2019) Information technology for peace and security. Springer Vieweg, Darmstadt

Rodicio-García ML, Ríos-de-Deus MP, Mosquera-González MJ, Abilleira MP (2020) The digital divide in Spanish students in the face of the COVID-19 crisis. Revista Internacional de Educacion para la Justicia Social 103–125

Salway RJ, Silvestri D, Wei EK, Bouton M (2020) Using information technology to improve COVID-19 care at New York City health+ hospitals: commentary describes New York City Health+ Hospitals use of innovative technological solutions to respond to the COVID-19 pandemic. Health Aff 39(9):1601–1604

Sanidas E (2004) Technology, technical and organizational innovations, economic and societal growth. Technol Soc 26(1):67–84

Seifert A (2020) The digital exclusion of older adults during the COVID-19 pandemic. J Gerontol Soc Work 63(6–7):674–676

Servon LJ (2008) Bridging the digital divide: technology, community and public policy. Wiley, New York

Seymour K, Skattebol J, Pook B (2020) Compounding education disengagement: COVID-19 lockdown, the digital divide and wrap-around services. J Children's Serv 15(4). https://doi.org/10.1108/JCS-08-2020-0049/full/html

Sherwin CW (1956) Securing peace through military technology. Bull Atomic Scientists 12(5):159–164

Shin H (2021) Governing the city through im/mobilities during COVID-19–technological self-governance. Eurasian Geography Econ, 1–15

Sholeh M, Ilhamuddin MF, Windasari W, Jacky M, Hafidz A (2020) Contact tracing model for positive Covid-19 patients with digital and local wisdom. In: International joint conference on science, technology, art, and humanities (IJCSTAH 2020), vol 1, No 1, pp 81–86

Smith A, Stirling A (2008) Social-ecological resilience and socio-technical transitions: critical issues for sustainability governance. Transitions. STEPS working paper 8, Brighton: STEPS Centre. Available from: https://opendocs.ids.ac.uk/opendocs/handle/20.500.12413/2438

Soete L, Turner R (1984) Technology diffusion and the rate of technical change. Econ J 94(375):612–623

Sostero M, Milasi S, Hurley J, Fernandez-Macias E, Bisello M (2020) Teleworkability and the COVID-19 crisis: a new digital divide? European Commission

Stauffacher D, Drake W, Currion P, Steinberger J (2005) Information and communication technology for peace: The role of ICT in preventing, responding to and recovering from conflict. United Nations/ICT Task Force

Stoneburner G (2001) Underlying technical models for information technology security. Available from: https://www.nist.gov/publications/underlying-technical-models-information-technology-security

Vogels E, Perrin A, Rainie L, Anderson M (2020) 53% of Americans Say the internet has been essential during the COVID-19 outbreak: Americans with lower incomes are particularly likely to have concerns related to the digital divide and the digital "homework gap'. Pew Research Center. Available from: https://eric.ed.gov/?id=ED609171

Wang T, Jia F (2021) The impact of health QR code system on older people in China during the COVID-19 outbreak. Age Ageing 50(1):55–56

Ware ND (2021) Online education in India and the widening digital divide. In: Handbook of Research on

Inequities in online education during global crises. IGI Global, pp 494–509
Warschauer M (2012) The digital divide and social inclusion. Am Q 6(2):131–135
Watts G (2020) COVID-19 and the digital divide in the UK. Lancet Digital Health 2(8):e395–e396
World Bank (2019) Myanmar country environmental analysis: a road towards sustainability, peace, and prosperity–synthesis report. World Bank
Wu J, Xie X, Yang L, Xu X, Cai Y, Wang T, Xie X (2021) Mobile health technology combats COVID-19 in China. J Infect 82(1):159–198
Yao Y, Zhang H, Liu X, Liu X, Chu T, Zeng Y (2021) Bridging the digital divide between old and young people in China : challenges and opportunities. Lancet Healthy Longevity 2(3):e125–e126
Zheng Y, Walsham G (2021) Inequality of what? An intersectional approach to digital inequality under Covid-19. Inform Organ 31(1):100341

19 Asset and Facilities Management from the Technical Perspective

We are always stronger with our inner power.
—Ali Cheshmehzangi, 2019

19.1 Knowing More About Asset and Facilities Management

Strategic asset management frameworks often respond to multiple sustainability factors, including peace and justice (Hanski and Ojanen 2020). In managing peace processes, we see growing demand for asset and facilities management and measures (Darby et al. 2000), including those that may be effective in the long run. For peace and stability, asset management plays a significant part in putting together practices that improve monitoring mechanisms, maintenance, and operation tasks, those that are commonly adopted from system operation thinking (Kezunovic 2004). Progressive approaches for developing asset management standards (Loistl and Petrag 2006) are examples of how asset management has been brought into relevant governance measures and technical sustainability motives. As a regulation factor, it could be discussed in terms of planning and budgeting, procurement processes, economic growth plans, and investment mechanisms (Simanjuntak et al. 2017). Hence, the move towards sustainability-based asset management (Marlow 2010; Marlow et al. 2010a) is essential for specific sectors, which means a more substantial commitment to broaden their focus (Marlow et al. 2010b). This point leads us to integrated asset management, which is considered an investment in sustainability (Henderson et al. 2014). It helps to move away from just the physical asset management to sustainability performance (Maletič et al. 2018), linking the two as well as exploring the relationships and signs of progress. These approaches suggest more comprehensive directions, such as sustainable asset performance, asset integrity management, and asset-facilities integration (Ratnayake and Liyanage 2009). The latter is of major interest to us from a technical sustainability perspective.

As seen in infrastructure asset management strategies (Shaw et al. 2015), there is a tangible scope to develop a system dynamics model (Mohammadifardi et al. 2019), reflecting on the technical demand of such strategies. Also related to the sustainable asset management area (Niekamp et al. 2015), we see solutions that entail traditional sustainability dimensions and correlate them with the technical dimension. In innovation-based studies, the move towards asset management services has generated newer debates on availability and sustainability factors (Ojanen et al. 2012), which help to improve physical asset performance and achieve a better asset lifecycle management (Schuman and Brent

A. Cheshmehzangi, *Urban Health, Sustainability, and Peace in the Day the World Stopped*, Sustainable Development Goals Series, https://doi.org/10.1007/978-981-16-4888-5_19

2005). In achieving these, we cannot neglect the importance of financial sustainability and technical integration of asset and facilities management in optimisation practices and research.

In this chapter, we intentionally put asset and facilities management together as we note the correlation between the two is more important than separating them in the health–sustainability–peace (HSP) studies. When it comes to asset and facilities, the combination of technical, socio-cultural, and financial aspects imply a significant understanding of sustainability assessment (Al-Sa'ed and Mubarak 2006). This also addresses some of the related technological development matters (Sadriddinov et al. 2020), including other management factors, such as production facilities, physical infrastructure, governance, etc. Moreover, we see the importance of investment strategy and project financing, which again require technical evaluation of various facilities and conducting risk assessment (Weber et al. 2016) and towards sustainable asset management. In doing so, we have to include specific sustainability metrics which will be used for managing what is available, how they will be used and managed, and where they can consider adaptive measures.

Focused on the topic from the technical viewpoint enables us to highlight the importance of both asset and facilities management in maintaining a good health–sustainability–peace nexus in cities and communities. Here, we focus on technical viewpoints and related factors that may be suitable for the better management and containment of future pandemics. Most of our discussions also reflect on socio-enviro-technical systems and sustainability matters (Pearce et al. 2010), which would help us consider asset life-cycle model, ownership management, and towards holistic framework development.

Examples like Multi-Criteria Prioritization Framework (MCPF) are suitable models that show how we could optimise the use of our infrastructure assets through better management (Elbarkouky 2012). This type of framework also responds to overarching sustainability issues of assets, particularly in developing community ownership and power in peace-making processes (Archer et al. 2019). Also partly related to asset integrity management, we see the importance of operational sustainability and various assets (Ratnayake 2010; Cheshmehzangi 2020a, b), which also addressed the shortfall and deficiencies of our cities during the COVID-19 pandemic. Thus, we delve into some of the key technical matters with the aim to optimise our asset and facilities management in cities and daily city operations.

19.2 Reflecting on How Cities Reacted During the Pandemic

During the COVID-19 pandemic, we recognise the multiple benefits of having robust asset and facilities management frameworks. The example of Ecuador's asset-based response plan helped the country to localise the needs of its healthcare facilities (Torres and Sacoto 2020), which were perhaps slightly late in responding to the impacts more effectively. The same applied when it came to economic support provisions, particularly for bank responses regarding lending facilities and asset purchases (Haas and Neely 2020). With growing demand in asset purchase and facilities use, cities had no option but to develop organisation resilience through asset management (Cheshmehzangi 2020a, c; Roberts 2020), which indeed helped to transform asset markets into capital markets (Ling et al. 2020; Sever et al. 2020) and stabilise the market in a relatively short period. As discussed by Caballero and Simsek (2020), the immediate shock by the pandemic was on non-financial matters, which indeed was on supply and demand matters. This shock effect had an impact on risk management matters, especially in sectors that depend on critical asset and regular facilities management processes. With emerging technologies to combat the pandemic, we see a surge in demands of critical care facilities, health units, and other related emergency and essential services (Vaishya et al. 2020). Earlier studies also highlighted the role of critical facilities in dealing with the pandemic crisis (Cheshmehzangi 2020a; Keenan 2020), which also means a better response plan through enhanced resilience and preparedness (Cheshmehzangi 2020c). We also note the technical importance of asset management processes (Marinelli 2020), through which physical infrastructure management would also be included.

In regard to facilities management, there is a wide range of examples that suggest the technical values behind optimising management processes. In particular, we see how mass testing facilities were utilised effectively through good use of public places, outdoor and indoor public environments and buildings, and adaptive use of various facilities. This model was successfully practised in cities across the UK (Peto 2020), ensuring minimal physical/social contact and effective use of time and space. An example of an agent-based model done by Cauevas (2020) shows us an evaluative technical approach to assess facilities and their risks on transmitting the disease. This could also be replicated for more severe conditions or more vulnerable facilities, such as outpatient healthcare facilities (Fisher et al. 2020; Wallace 2020), nursing homes residents (Su et al. 2020), childcare facilities (Lopez et al. 2020), and long-term care facilities (LTCF), where higher transmission risks are expected (Danis et al. 2020; Heudorf et al. 2020; Iritani et al. 2020; Sarah et al. 2020; WHO 2020a). Many studies of the former example are based in the US healthcare facilities, also showing the effectiveness of technical support for asset and facilities management during the pandemic.

During the pandemic, we note a significant emphasis was on connecting urban assets together, ensuring they are managed promptly to enable better flows in cities (Cheshmehzangi 2020a; also see 100 Resilient cities 2018). From the techno-operational perspective, we see many examples of adaptive measures and planning, which also reflect on the flexibility in urban assets and available facilities. On the other hand, however, we witnessed many examples that lacked quality management, especially regarding temporary repurposing or reuses, as well as readjustments in asset and facilities management processes. Thus, we note the two are inseparable, indicating why we need to consider integration and technical support rather than the mere availability of assets, facilities, and supplies.

In some cases, we witnessed failures in preventing the disease infection. For instance, there was a lack of control compliance in outpatient facilities in Tanzania (Powell-Jackson et al. 2020), which was the result of neglect at the governmental level. Preparedness of frontline doctors was indeed recognised as the right approach to battle the pandemic in the worst situations. Still, scholars like Suleiman et al. (2020) noted serious threats to healthcare systems in countries like Jordan. The quality of facilities was also questioned regarding the effectiveness of control and management scenarios, such as those in nursing homes/facilities in California (He et al. 2020). The collapse of medical facilities combined with lockdown pressures (Sugishita et al. 2020) was not a case of one city or one country. Still, it immediately became a common scenario of cities that were able to manage their facilities in accordance with high-level safety measures. The situation with significant economic challenges, such as the case of Iran, was slightly different in mobilising government facilities, community, and civil society (Zandifar and Badrfam 2020). Nonetheless, with inequalities between countries in the European Union (EU), we witness similar issues regarding the relocation of certain facilities and healthcare units to certain regions at the time of need. The Italian Lombardi region is an example, which became an immediate European centre (Cheshmehzangi et al. 2021), mainly because of lack of facilities support in the first phase of the pandemic.

The success stories also include examples of mass testing facilities for vulnerable groups and units (Louie et al. 2020), which helped speed up testing processes in cities across the USA. Other cases included methods of dealing with medical and healthcare wastes in the original epicentre, the City of Wuhan, China (Yang et al. 2021), or propositions to develop immediate quarantine measures and facilities (Li et al. 2020), such as those also experienced in Pakistan (Waris et al.

2020) and Italian cities (Remuzzi and Remuzzi 2020). The built environment management seemed very tricky at first, under many uncertainties and lack of knowledge about the disease. Still, it soon became common to prevent and control the pandemic in vulnerable facilities and units (Wang 2021). In China, we also see the effectiveness of neighbourhood-level public facilities (Jin et al. 2020) or adaptive measures for repurposing communal and community environments (Cheshmehzangi 2020a, e). In the process of flattening the curve, we noticed the effectiveness of adaptive measures in providing necessary facilities, such as those that took into consideration the regional differences and community inequalities. In doing so, city governments could relocate assets and facilities when and where needed and use technical experts to guide specific sectors. This technical guidance approach (WHO 2020a) is valued as part of developing quarantine centres, safety checkpoints, mass testing facilities, vaccination centres, etc. These examples would enable us to consider three aspects of critical preparedness, readiness, and response (Cheshmehzangi 2020a; WHO 2020b), which are essential for asset and facilities management, too.

19.3 Dealing with Shortfalls and Shortages in Cities

As the pandemic advances in many countries, we see a growing demand for technical guidance support of cities and specific sectors (WHO 2020a). We also see an urgent need to build a proactive engagement culture in organisations that include asset and facilities management (Brunetto et al. 2014). To do so, we could focus on the sustainability and longevity of our assets and facilities (ibid), including attention on technical aspects relevant to shortfalls and shortages in cities. The pandemic did highlight not only shortage issues but also deficiencies and lack of technical support in managing the outbreaks or other emergencies. In larger cities, we see the complications could lead to inefficient use of assets and facilities. And in small- and medium-sized cities, we see a lack of facilities to start with, which could complicate the situation if adaptive measures are not in place.

As highlighted before (Cheshmehzangi 2020a), we stress the importance of adaptive thinking in optimising city operations and their management. We see the struggle is related to the lack of technical integration of such approaches, which (again) should be carefully looked at for future pandemics. Thus, we argue the shortfalls and shortages are based on four interconnected areas. The first is related to asset and facilities shortages. The second is related to shortfalls in management processes, such as in facilities management. The third is related to lack of technical support, leading to weaker policies and the decision-making process. Finally, the fourth is related to the absence of adaptive planning, which is evident in many cases across the globe.

19.4 A Summary

As the COVID-19 pandemic hit, cities were faced with immediate and long-term shortages and shortfall. Much of our issues are related to corrupted institutions or sectors that have an inability to manage the situation promptly. Perhaps this is meant to be intentional to show inactions and poor management situations, similar to what could also be the cause of prolonged construction projects that take years in places where facilities, assets, and workers are widely available. Again, perhaps intentional, as there is a continuing lack of technical and financial support in maintaining city operations, enhancing the adaptive measures, and ensuring high-quality management is utilised for assets and facilities.

This chapter sheds light on some of the critical issues related to our asset and facilities management from a technical perspective. We explored some failure and success cases, showing how cities reacted and adapted new measures and practices. The one factor that we did not discuss enough is related to policies and regulations, which must be embedded in decision-making processes for future asset and facilities management plans. We also see opportunities for technical solutions regarding creating smart platforms, not necessarily for the smartness of cities, but to achieve smart-resilient models for cities and communities. This will be the focus of our next chapter, where we aim to bring together the technological–technical nexus with the increasing smart platforms around the globe. The two can help us achieve new integrated urban models, which will be valuable to reach adequate urban health, enhance sustainability, and maintain peace.

References

100 Resilient Cities (2018) What is urban resilience? Available from: https://www.100resilientcities.org/resources/

Al-Sa'ed R, Mubarak S (2006) Sustainability assessment of onsite sanitation facilities in Ramallah-Albireh district with emphasis on technical, socio-cultural and financial aspects. Manage Environ Quality: Int J 17(2):140–156

Archer T, Batty E, Harris C, Parks S, Wilson I, Aiken M et al (2019) Our assets, our future: the economics, outcomes and sustainability of assets in community ownership. Power to Change and Ministry of Housing, Communities and Local Government. Technical Report, 30 p, Published in July 2019, Available from: https://www.powertochange.org.uk/wp-content/uploads/2019/07/Our-assets-our-future_Technical-Report.pdf

Brunetto Y, Xerri M, Nelson S (2014) Building a proactive, engagement culture in asset management organizations. J Manag Eng 30(4):04014014

Caballero RJ, Simsek A (2020) Asset prices and aggregate demand in a "Covid-19" shock: a model of endogenous risk intolerance and LSAPS. National Bureau of Economic Research. NBER Working Paper 27044, Available from: https://www.nber.org/system/files/working_papers/w27044/revisions/w27044.rev2.pdf

Cheshmehzangi A (2020a) The city in need: urban resilience and city management during disruptive disease outbreak events. Springer, Singapore

Cheshmehzangi A (2020b) Reflection on disruptions: managing the city in need, saving the city in need. In: Cheshmehzangi A (ed) The city in need: urban resilience and city management during disruptive disease outbreak events. Springer, Singapore, pp 137–283

Cheshmehzangi A (2020c) Preparedness through urban resilience. In: Cheshmehzangi A (ed) The city in need: urban resilience and city management during disruptive disease outbreak events. Springer, Singapore, pp 41–104

Cheshmehzangi A (2020d) Reflection on early lessons for urban resilience and public health enhancement during the COVID-19. Health 12(10):1390

Cheshmehzangi A (2020e) 10 Adaptive measures for public places to face the COVID 19 pandemic outbreak. City Soc 32(2)

Cheshmehzangi A, Sedrez M, Ren J, Kong D, Shen Y, Bao S, Xu J, Su Z, Dawodu A (2021) The effect of mobility on the spread of COVID-19 in light of regional differences in the European Union. Sustainability 13(10):5395. https://doi.org/10.3390/su13105395

Cuevas E (2020) An agent-based model to evaluate the COVID-19 transmission risks in facilities. Comput Biol Med 121:103827

Danis K, Fonteneau L, Georges S, Daniau C, Bernard-Stoecklin S, Domegan L et al (2020) High impact of COVID-19 in long-term care facilities, suggestion for monitoring in the EU/EEA, May 2020. Eurosurveillance 25(22):2000956

Darby J, Mac Ginty R, Mac Ginty R (eds) (2000) The management of peace processes. Springer, Cham

Elbarkouky MM (2012) A multi-criteria prioritization framework (MCPF) to assess infrastructure sustainability objectives and prioritize damaged infrastructure assets in developing countries. J Sustain Dev 5(9):1

Fisher KA, Tenforde MW, Feldstein LR, Lindsell CJ, Shapiro NI, Files DC, Gibbs KW, Erickson HL, Prekker ME, Steingrub JS, Exline MC (2020) Community and close contact exposures associated with COVID-19 among symptomatic adults ≥ 18 years in 11 outpatient health care facilities—United States, July 2020. Morb Mortal Wkly Rep 69(36):1258

Haas J, Neely CJ (2020) Central Bank responses to COVID-19. Econ Synopses (23)

Hanski J, Ojanen V (2020) Sustainability in strategic asset management frameworks: a systematic literature review. Int J Strategic Eng Asset Manage 3(4):263–294

He M, Li Y, Fang F (2020) Is there a link between nursing home reported quality and COVID-19 cases? Evidence from California skilled nursing facilities. J Am Med Dir Assoc 21(7):905–908

Heudorf U, Müller M, Schmehl C, Gasteyer S, Steul K (2020) COVID-19 in long-term care facilities in Frankfurt am Main, Germany: incidence, case reports, and lessons learned. GMS Hygiene Infection Control 15

Iritani O, Okuno T, Hama D, Kane A, Kodera K, Morigaki K et al (2020) Clusters of COVID-19 in long-term care hospitals and facilities in Japan from 16 January to 9 May 2020. Geriatr Gerontol Int 20 (7):715–719

Jin X, Leng Y, Gong E, Xiong S, Yao Y, Vedanthan R et al (2020) Neighborhood-level public facilities and COVID-19 transmission: a nationwide geospatial study in China

Keenan JM (2020) COVID, resilience, and the built environment. Environ Syst Decis 1-6. https://doi.org/10.1007/s10669-020-09773-0

Kezunovic M (2004) Integrating data and sharing information from various IEDs to improve monitoring, condition-based diagnostic, maintenance, asset management, and operation tasks. In: EPRI substation equipment disturbance conference, February 2004. Available from: http://smartgridcenter.tamu.edu/resume/pdf/cnf/EPRI04.pdf

Li J, Yang Y, Gong M, Shi J, Zhou X, Xing X et al (2020) Aggressive quarantine measures reduce the high morbidity of COVID-19 in patients on maintenance hemodialysis and medical staff of hemodialysis facilities in Wuhan, China. Kidney Diseases 6(4):271–283

Ling DC, Wang C, Zhou T (2020) A first look at the impact of COVID-19 on commercial real estate prices: asset-level evidence. Rev Asset Pricing Stud 10 (4):669–704

Loistl O, Petrag R (2006) Asset management standards: corporate governance for asset management. Springer, Cham

Lopez AS, Hill M, Antezano J, Vilven D, Rutner T, Bogdanow L et al (2020) Transmission dynamics of COVID-19 outbreaks associated with child care facilities—Salt Lake City, Utah, April–July 2020. Morb Mortal Wkly Rep 69(37):1319

Louie JK, Scott HM, DuBois A, Sturtz N, Lu W, Stoltey J et al (2020) Lessons from mass-testing for COVID-19 in long term care facilities for the elderly in San Francisco. Clin Infect Dis. Ciaa1020. https://doi.org/10.1093/cid/ciaa1020/5873783

Marlow DR (2010) Sustainability-based asset management in the water sector. In: Definitions, concepts and scope of engineering asset management. Springer, London, pp 261–275

Marlow DR, Beale DJ, Burn S (2010a) A pathway to a more sustainable water sector: sustainability-based asset management. Water Sci Technol 61(5):1245–1255

Marlow DR, Beale D, Burn S (2010b) Linking asset management with sustainability: views from the Australian sector. J Am Water Works Ass 102 (1):56–67

Maletič D, Maletič M, Al-Najjar B, Gomišček B (2018) Development of a model linking physical asset management to sustainability performance: an empirical research. Sustainability 10(12):4759

Mohammadifardi H, Knight MA, Unger AA (2019) Sustainability assessment of asset management decisions for wastewater infrastructure systems—Implementation of a system dynamics model. Systems 7 (3):34

Niekamp S, Bharadwaj UR, Sadhukhan J, Chryssanthopoulos MK (2015) A multi-criteria decision support framework for sustainable asset management and challenges in its application. J Ind Prod Eng 32(1): 23–36

Ojanen V, Ahonen T, Reunanen M, Hanski J (2012) Towards availability and sustainability in customer value assessment of asset management services. Int J Innov Sustain Dev 6(4):368–391

Pearce AR, Bernhardt KLS, Garvin MJ (2010, December) Sustainability and socio-enviro-technical systems: Modeling total cost of ownership in capital facilities. In: Proceedings of the 2010 winter simulation conference. IEEE, New York, pp 3157–3169

Peto J (2020) Covid-19 mass testing facilities could end the epidemic rapidly. BMJ 368

Powell-Jackson T, King JJ, Makungu C, Spieker N, Woodd S, Risha P, Goodman C (2020)Infection prevention and control compliance in Tanzanian outpatient facilities: a cross-sectional study with implications for the control of COVID-19. Lancet Global Health 8(6):e780–e789

Ratnayake RMC (2010) Asset integrity management: operationalizing sustainability concerns. Handbook on business information systems. World Scientific Publishing Co, Hackensack, pp 323–357

Ratnayake RMC, Liyanage JP (2009) Asset integrity management: sustainability in action. Int J Sustain Strategic Manage 1(2):175–203

Remuzzi A, Remuzzi G (2020) COVID-19 and Italy: what next? The Lancet 395(10231):1225–1228

Roberts C (2020) Developing organization resilience through asset management to respond to COVID-19. IAM Responding to COVID-19, 5 pages, Available from: https://theiam.org/media/2358/developing-organization-resilience-through-asset-management-to-respond-to-covid-19.pdf

Sadriddinov MI, Mezina TV, Morkovkin DE, Romanova JA, Gibadullin AA (2020) Assessment of technological development and economic sustainability of domestic industry in modern conditions. In: IOP conference series: materials science and engineering, vol 734, No 1, p 012051). IOP Publishing

Sarah HY, See I, Kent AG, Vlachos N, Whitworth JC, Xu K et al (2020) Characterization of COVID-19 in assisted living facilities—39 states, October 2020. Morb Mortal Wkly Rep 69(46):1730

Schuman CA, Brent AC (2005) Asset life cycle management: towards improving physical asset performance in the process industry. Int J Oper Prod Manage 25(6). https://doi.org/10.1108/01443570510599728/full/html

Sever C, Goel R, Drakopoulos D, Papageorgiou E (2020) Effects of emerging market asset purchase program announcements on financial markets during the COVID-19 pandemic. IMF Working Paper no. 2020/292, 22 pages, Available from: https://papers.ssrn.com/sol3/papers.cfm?abstract_id=3772498

Shaw G, Walters R, Kumar A, Sprigg A (2015) Sustainability in infrastructure asset management. In: Proceedings of the 7th World Congress on engineering asset management (WCEAM 2012). Springer, Cham, pp 525–534

Simanjuntak J, Ratnawati T, Rahmiyati N (2017) Economic growth as mediation of regional own source revenue, investment and asset management on labor absorption and social welfare in regencies/cities in Riau Islands Province. Int J Econ Fin 9(8):127

Su Z, Meyer K, Li Y, McDonnell D, Joseph NM, Li X et al (2020) Technology-based interventions for nursing home residents: implications for nursing home practice amid and beyond the influence of COVID-19: a systematic review protocol. Pre-print paper. Available from Research Square, at: https://assets.researchsquare.com/files/rs-56102/v2/ae8d73b9-373b-49b1-b323-11e68981dfd7.pdf

Sugishita Y, Kurita J, Sugawara T, Ohkusa Y (2020) Forecast of the COVID-19 outbreak, collapse of medical facilities, and lockdown effects in Tokyo, Japan. medRxiv

Suleiman A, Bsisu I, Guzu H, Santarisi A, Alsatari M, Abbad A et al (2020) Preparedness of frontline doctors in Jordan healthcare facilities to COVID-19 outbreak. Int J Environ Res Public Health 17(9):3181

Torres I, Sacoto F (2020) Localising an asset-based COVID-19 response in Ecuador. The Lancet 395 (10233):1339

Vaishya R, Haleem A, Vaish A, Javaid M (2020) Emerging technologies to combat the COVID-19 pandemic. J Clin Exp Hepatol 10(4):409–411

Wallace M (2020) Public health response to COVID-19 cases in correctional and detention facilities—Louisiana, March–April 2020. MMWR. Morbidity and mortality weekly report 69

Wang Z (2021) Use the environment to prevent and control COVID-19 in senior-living facilities: an analysis of the guidelines used in China. HERD: Health Environ Res Des J 14(1):130–140

Waris A, Khan AU, Ali M, Ali A, Baset A (2020) COVID-19 outbreak: current scenario of Pakistan. New Microbes and New Infections, 100681

Weber B, Staub-Bisang M, Alfen HW (2016) Infrastructure as an asset class: investment strategy, sustainability, project finance and PPP. Wiley, New York

WHO (World Health Organization) (2020a) Infection prevention and control guidance for long-term care facilities in the context of COVID-19: interim guidance, 21 March 2020 (No. WHO/2019-nCoV/IPC_long_term_care/2020.1). World Health Organization

WHO (World Health Organization) (2020b) Critical preparedness, readiness and response actions for COVID-19: interim guidance, 22 March 2020 (No. WHO/2019-nCoV/Community_Actions/2020.3). World Health Organization

Yang L, Yu X, Wu X, Wang J, Yan X, Jiang S, Chen Z (2021) Emergency response to the explosive growth of health care wastes during COVID-19 pandemic in Wuhan, China. Resources Conserv Recycl 164:105074

Zandifar A, Badrfam R (2020) Fighting COVID-19 in Iran; economic challenges ahead. Arch Iran Med 23 (4):284

20 Smart Platforms and Technical Solutions: Can We Really Achieve Smart-Resilient Models?

We are not so smart as beings. We just learned how to educate ourselves long ago. So now we should be worried, as we are focused on educating the machines.

—Ali Cheshmehzangi, 2020

20.1 Knowing More About Smart Platforms and Technical Solutions

In urban literature, we often see resilient and smart together with other terms such as the sustainable circular economy (Van Fan et al. 2019), smart technologies (Bellini and Nesi 2018), climate-sensitive urban development (Galderisi and Colucci 2018), Internet of things (IoT) (Lazaroiu and Roscia 2018), ICTs (Baron 2012), etc. We often develop a set of indicators (Khatibi et al. 2021) to help design and planning processes of smart-resilient urban models. In a way, smart-resilient (as well as smart and resilient) models provide the opportunity to implement technical solutions and support governance of cities (Van Fan et al. 2019), which help to speed up sustainable transitions (Galderisi and Colucci 2018; Cheshmehzangi and Chen 2021; Cheshmehzangi et al. 2021a). In many cases, we see a lack of technical and financial capacity (McLeod et al. 2019) that becomes the barrier to smart-resilient pathways, which should be addressed to ensure innovative approaches are developed and pushed forward. For instance, an ICT service enablement (Baron 2012) could help us utilise "*sensors and actuator deployments everywhere*", which could then be relevant to smart-resilient strategies, helping redefine the city's technological aspect.

Examples of socio-technical approaches focus on resilient connections and socio-technical systems (STSs), such as in smart transportation systems (Roy et al., 2021) or in other smart-resilient practices such as in climate-resilient agriculture (Prasad et al. 2014). The emergence of new technologies and computing networks has led to a series of recent technical problems, such as network problems (Doan et al. 2018; Liu et al. 2019). However, the shift towards automation and computational optimisation processes has opened up new discussions on smart-resilient practices. By questioning if smart cities are resilient, scholars have looked into how smartness could lead towards resilience thinking (Zhu et al. 2019), if not the other way around. The opportunities for hybrid models (Cheshmehzangi 2017) are widespread, allowing for new paradigms, practices, and towards experimental projects to achieve sustainable development transitions (Tan-Mullins et al. 2017; Xie et al. 2019). The integration of resilient networking in smart grids (Kher et al. 2013; Tsado et al. 2014; Wu et al. 2015; Cheshmehzangi et al. 2021a) or cross-sectoral strategies by combining the two

A. Cheshmehzangi, *Urban Health, Sustainability, and Peace in the Day the World Stopped*, Sustainable Development Goals Series, https://doi.org/10.1007/978-981-16-4888-5_20

approaches (Papa et al. 2015) help to verify (if not develop) the values of smart-resilient thinking. This means that we could provide pathways to ensure our resilience is smarter, enabling new practices that are efficient, effective, and responsive.

New approaches like the use of the blockchain-based hybrid network in smart cities (Sharma and Park 2018) help us to include a resilient hybrid model in smart systems and models of development. Other examples entail management optimisation, the use of new approaches of machine learning for IoT-enabled urban systems (Babar et al. 2020), and some suggest directions towards the promotion of e-governance or e-government in the current digital age. The idea is not necessarily new, as it started with the overarching arguments of e-governance for smart communities almost two decades ago (Coe et al. 2001; Sealy 2003). Yet, the new integration of ICTs and blockchain in e-governance has helped to enhance the idea of smart cities (Oliveira et al. 2020), and perhaps from the resilient thinking approach, too. In Europe, the progress of city e-governance emerged as a method to enable smart city practices and enhance the decision-making process through human capital and democratic activities (Paskaleva 2009). While the concept remains quite broad (Vinod Kumar 2015), we see many enabling factors that may be useful for future smart-resilient directions. This requires system thinking approaches or at least specific technical measures that could enhance the use of ICTs in smart-resilient practices (Alotaibi 2018; Cortés-Cediel et al. 2017; Tang et al. 2019). For instance, we see a greater opportunity for resource efficiency of any sort, which is often not achieved through mere smart practices. The use of ICTs for integrated governance (Kumar 2017) also suggests ways of efficient service provision, faster responsiveness or reaction to situations, and "*citizen empowerment through access to information*". While we doubt the latter point, we see efficiency and effectiveness as major aspects that require careful attention in resilience enhancement in city management practices (Cheshmehzangi 2020a, b). Thus, we see a growing demand for citizen participation, particularly with the rise of digital media platforms (Gil et al. 2019), in achieving e-governance in perhaps towards smart-resilient directions.

This chapter is the closing chapter of all aspects studies in six parts of the book. It finishes the discussions of technical dimension and suggests smart-resilient models achieve a sound health–sustainability–peace (HSP) nexus in cities and communities. It discusses related factors to smart platforms and technical solutions, which is believed to be another crossover between technical and technological directions. This chapter explicitly explores the topic of smart-resilience and suggests ways in which it could be achieved in practice. This aspect is important to future planning and governance pathways and those directions that entail policy-making and integrated action plans. The discussions also partly include governance and social factors, which are again important in developing healthy smart-resilient practices for cities and communities. The use of ICTs and other emerging technological instruments are discussed here to ensure meeting smart urban management scenarios and resilient directions for a better future. These discussions, however, should be considered cautiously as the growing digital world that we currently experience may not necessarily be the best direction of all time. Indeed, we may need to question and redefine our smartness and resilience and enhance governance and technological–technical opportunities for a healthier future. Thus, the chapter can only suggest some viewpoints in regard to the use of smart systems, technologies, and platforms as technical solutions for impending healthy, sustainable, and peaceful societies.

20.2 Reflecting on How Cities Reacted During the Pandemic

In research and partly in practice, significant attention has been on big data and artificial intelligence (AI) integration in managing the

COVID-19 pandemic (Bragazzi et al. 2020). In this regard, smart pandemic management has been suggested as an approach to smart, resilient, and flexible decision-making systems (Mezzour et al. 2020). The smart-resilient strategy seems to be the major focus on helping with demand–supply issues (Di Tommaso 2020) and other management factors (Cheshmehzangi 2020a), as well as transformative approaches that entail system thinking and big data use (Cheshmehzangi 2021). From smart-resilient industrial development (Di Tommaso 2020) to e-governance pathways (Ullah et al. 2021), we see traces of promoting sustainable development and smart-resilient measures. Thus, we need to consider excellence in management processes, create accessible services, and rethink public governance (Ata-Agboni and Olufemi 2021). We may need to see the values behind the shifts for the latter, driven in particular from deficiencies of current management and governance practices.

Many scholars have appraised and recommended the use of ICTs during and after the COVID-19 pandemic. For instance, Arshad (2020) suggests, "*the role of Information and Communication Technology (ICT) is significant amid the disease outbreaks and the new normal. Whereby resilience is a key factor initiated by the business and risk leaders in a world where the new normal is the disease outbreaks*". The others argue the emergence of ICT in specific sectors, such as education in the post-COVID-19 era (Naresh 2020; Adarkwah 2021). The role of ICT and social media was also discussed as a breakthrough strategy that helped the continuity of business operations and working environments (Saleh 2020; Su et al. 2021). Although in different ways, the integration of ICT in existing practices helped reduce the pandemic's economic impacts slightly. However, we note this is a contextual matter, which becomes a completely different scenario in poorer communities or countries. ICT intervention in the containment of the COVID-19 pandemic, as suggested by Zaman et al. (2020), helped develop smart-resilient practices and become immediate solutions to effective management (Strielkowski et al., 2021). In cases like in Italian cities, we see the role of ICT-based solutions in crisis and emergency plans (Tropea and De Rango 2020). We also see the effectiveness of ICT platforms for public health systems (Lorente et al. 2020), mobility pattern analysis, and other factors that seem critical in achieving decent governance during the pandemic.

As discussed by Zheng and Walsham (2021) and in Chap. 18 of the book, digital inequality under the COVID-19 pandemic could lead to complex long-term issues. This matter has been experienced in communications during physical distancing time (Nguyen et al. 2021), remote teaching and learning platforms (Czerniewicz et al. 2020), and healthcare support. For instance, while ICT has been effective in schools and the education sector in Japan (DeWit et al. 2020), it has not stopped the emerging educational inequalities (Sato and Saito 2020). Nonetheless, the situation has led to new movements to reshape education services (Kang 2021), which could happen through improvements in ICT infrastructure and better education services across the country. The same scenario was experienced in healthcare services, such as during the testing and vaccination stage, where there was criticism against the exclusive use of ICT-mediated platforms. Such an approach led to disengagement with more impoverished communities, the elderly, vulnerable groups, and those with little digital knowledge. Some even suggest that digital inequalities themselves have become a new form of vulnerability (Robinson et al. 2020), which does not surprise us at all.

In facing the digital divide and economic inequalities, cities in Italy started to focus on value governance (Marino and Pariso 2020). As Yang (2020) suggests, "*COVID-19 manifested the pivotal role of public value governance*", and therefore, we see the need to better balance economy and human life (ibid). The shift towards e-governance has been the result of such thinking, which also suggests improvements in service delivery (Bhuvana and Vasantha 2020), public communication, social media use (Bosman 2021), and towards COVID-19-induced digital acceleration (Agostino et al. 2021). In the South African case, Bosman (2021, p. 4) suggests some significant changes to the emerging e-government model:

> Increasing mobile phone penetration on the continent not only enables Africans to access social media but is also useful for 'maintaining networks of family and friends, for mobile banking, for comparing market prices, for collecting health data, for advertising and for finding jobs'. Some governments have been using these technologies in innovative ways in their pandemic responses. South Africa, in particular, stands out. Its existing platforms of e-government, as well as new ways of relying on social media and smartphones in its pandemic response, could hold valuable lessons for future e-government strategies in the rest of Africa.

Thus, we see the merge between government and social media to become a fast-growing approach to push e-governance in specific contexts. As a technical solution, this could help democratisation processes in smart-resilient cities, but it could also lead to conflicts that may not be so welcomed in every country. As suggested by Song and Lee (2016, p. 433), e-government practices are recognised as "*the use of the internet and other digital media to deliver government information and services to citizens*'. Thus, this could only be a contextual matter as it crosses over countries and regions' political, social, and cultural factors. At the city level, however, e-governance could help promote public and private values in citizen partnership (Ju et al. 2019), which also indicate the flexibility of processes, and healthy dynamism (Singh et al. 2021) for dealing with challenges that exist when connecting people to smart cities (Oliveira et al. 2020). Some also argue that e-governance could help to bridge the digital divide (Stoiciu 2011), which we argue that could only occur during the non-emergency times.

20.3 Achieving Smart-Resilient Cities and Communities

As we learnt from the COVID-19 pandemic, information technologies cannot be simply used for detection processes. They have to be utilised for enhancing the smart-resilience of communities. However, the question is can we have such models in practice? Yes is the short answer if we consider integrated models that are comprised of flexibility, adaptability, and the human dimension. We see smart platforms becoming technical solutions to optimise performances, management, and processes. Thus, common practices may not be the limit, yet we insist on the inclusion of human-centric measures and indicators.

This chapter highlighted some of the key aspects of smart-resilient direction and/or practices. Other common aspects of smart directions and technical solutions mainly focus on sensory network developments in specific systems, such as medical and healthcare systems (Ali et al. 2020). The future connectedness aspect, such as the Internet of Things (IoT) (Gupta et al. 2021), is also widely recognised as an approach to help to make communities smarter and resilient, especially when facing calamities. The virtual connectivity enables us to develop intelligent solutions for higher risk environments, which is discussed by Dolgikh (2020) when it comes to technological and policy solutions in risk management, as well as for the development of risk management frameworks in communities and cities. The IoT-based approaches were commonly experienced in cities, which helped to fast-forward the development of detection and diagnosis system (Mohammed et al. 2020) and data-driven responses (James et al. 2020). Such approaches also opened up new opportunities for the development of smart platforms. We name a few here, including the use of smart city technologies (Inn 2020), smart material applications (Javaid and Haleem 2020), smart and sustainable Nano-technological solutions (Jastrzębska and Vasilchenko 2021), smart materials-integrated sensor technologies (Erdem et al. 2021), real-time detection through a network of smart thermometers (Chamberlain et al. 2020), time-space surveillance technology (Sonn and Lee 2020), AI and facial recognition technologies (Cheshmehzangi 2020a), neuroscience (Riva et al. 2021), etc.

Lastly, we reflect on a wide range of smart solutions using smart city technology, which is discussed thoroughly by Jaiswal et al. (2020). In this regard, we see remote solutions were practiced widely but would not necessarily solve future operations. This, we have to be cautious about transformative shifts that may simply become ubiquitous practices with carefully crafted smart city initiatives, big data use, robotic use, and AI integration in most practices. The gradual progress is perceptible, yet we must understand the values behind "smart-resilience' and not 'smart alone". This is why we see only a handful examples of genuine smart-resilient practices, which focused on not abusing the data but truly using them to enhance the resilience of cities and communities. After all, we believe that smartness alone, particularly based on technological advancements, would be perilous to humanity and its decisive survival.

20.4 A Summary

We see a wide range of smart platforms and technical solutions, specifically to promote smart-resilient strategies, models, and development. The COVID-19 pandemic and its longer-term challenges have fast-forwarded many smart and resilient initiatives in cities and communities. From IoT systems to data-driven monitory platforms, we see the values of technical–technological solutions offered for smart-resilient directions. In the wake of social media and the e-governance era, we see opportunities for peace-building processes, including people in decision-making processes and ensuring healthier public participation.

As we close our explorations of all aspects related to six sustainability dimensions, we delve into further debates of smart-resilient models and practices, which we believe are relevant for future HSP indicators. To have a healthy society, we could invest in specific technical solutions, such as smart e-health systems (Khan and Karim 2020) and other smart technology adoption that could enhance our healthcare operations, maintain health in communities, and create innovative solutions while facing our current deficiencies and shortfalls. We see a growing opportunity for smart technologies driven approaches (Khan et al. 2021), data-driven analytical methods (Cheshmehzangi et al. 2021b), and holistic interventions that include smart and resilient indicators. However, the future remains uncertain. The growing demand for AI integration, robots, and similar technologies, may seem to overlay the genuine smart-resilient practices offered by current practices of IoT system use, Internet+, ICTs, etc. If we opt to move towards upgrading our systems through information-based platforms, we may encounter a wide range of inequalities and challenges that may not seem so effective for all. To implement an all-inclusive approach, we need to use or utilise and not rely on smart technologies. The more we depend on such technologies, the more we may move away from our smartness as collective beings. Individually, we may succeed in the long run, but collectively, we will fail.

The next chapter is the concluding chapter of this book. We have already covered 18 factors across six overarching dimensions of physical, environmental, economic, social, institutional, and technical. It is time for us to conclude on the topic of the city we need in facing pandemics, emergencies, and other calamities. The narrative throughout the book and all aspects has been centred around the HSP nexus in cities and communities, and we continue to debate this topic as we note its importance in achieving a healthier future.

References

Adarkwah MA (2021) "I'm not against online teaching, but what about us?": ICT in Ghana post Covid-19. Educ Inf Technol 26(2):1665–1685

Agostino D, Arnaboldi M, Diaz L (2021) New development: COVID-19 as an accelerator of digital transformation in public service delivery. Public Money Manage Taylor Francis J 41(1):69–72. Available from: https://ideas.repec.org/a/taf/pubmmg/v41y2021i1p69-72.html

Ali S, Singh RP, Javaid M, Haleem A, Pasricha H, Suman R, Karloopia J (2020) A review of the role of smart wireless medical sensor network in COVID-19. J Indus Integr Manage 5(4)

Alotaibi SS (2018) Registration center based user authentication scheme for smart e-governance applications in smart cities. IEEE Access 7:5819–5833

Arshad M (2020) COVID-19: it's time to be thankful to our ICT professionals. Inf Technol Electr Eng 9 (2):23–31

Ata-Agboni JU, Olufemi IO (2021) E-Governance and E-Government: rethinking public governance in Nigeria, within the context of COVID-19. J Good Gover Sustai Dev Africa 6(2):54–59

Babar M, Tariq MU, Jan MA (2020) Secure and resilient demand side management engine using machine learning for IoT-enabled smart grid. Sustain Cities Soc 62:102370

Baron M (2012) Do we need smart cities for resilience? J Econ Manage 10:32–46

Bellini E, Nesi P (2018) Exploiting smart technologies to build smart resilient cities. Routledge handbook of sustainable and resilient infrastructure. Routledge, New York and London, pp 685–705

Bhuvana M, Vasantha S (2020) Rural citizen satisfaction on e-Health care services under e-Governance service delivery model during COVID 19. Int J Manage 11 (09)

Bosman I (2021) COVID-19 and E-governance: lessons from South Africa. 18 pages report, Available from: https://www.jstor.org/stable/pdf/resrep29596.pdf

Bragazzi NL, Dai H, Damiani G, Behzadifar M, Martini M, Wu J (2020) How big data and artificial intelligence can help better manage the COVID-19 pandemic. Int J Environ Res Public Health 17(9):3176

Chamberlain SD, Singh I, Ariza CA, Daitch AL, Philips PB, Dalziel BD (2020) Real-time detection of COVID-19 epicenters within the United States using a network of smart thermometers. medRxiv. Available from: https://www.medrxiv.org/content/10.1101/2020.04.06.20039909v1

Cheshmehzangi A (2017) Sustainable development directions in china: smart, eco, or both? In: The 2017 international symposium on sustainable smart eco-city planning and development, held in Cardiff, Wales, pp 11–13

Cheshmehzangi A (2020) The city in need: urban resilience and city management during disruptive disease outbreak events. Springer, Singapore

Cheshmehzangi A (2020b) Reflection on disruptions: managing the city in need, saving the city in need. In: Cheshmehzangi A (ed), The city in need: urban resilience and city management during disruptive disease outbreak events. Springer, Singapore, pp 137–283

Cheshmehzangi A (2021) From transitions to transformation: A brief review of the potential impacts on COVID-19 on boosting digitization, digitalisation, and systems thinking in the built environment. J Build Construct Plann Res 9(1):107749:1–14

Cheshmehzangi A, Chen H (2021) China's sustainability transitions: low carbon and climate-resilient plan for carbon neutral 2060. Springer, Singapore

Cheshmehzangi A, Dawodu A, Sharifi A (2021a) Sustainable urbanism in China. Routledge, New York and London

Cheshmehzangi A, Sedrez M, Ren J, Kong D, Shen Y, Bao S, et al (2021b) The Effect of mobility on the spread of COVID-19 in light of regional differences in the European Union. Sustainability 13:5395

Coe A, Paquet G, Roy J (2001) E-governance and smart communities: a social learning challenge. Soc Sci Comput Rev 19(1):80–93

Cortés-Cediel ME, Cantador I, Gil O (2017) Recommender systems for e-governance in smart cities: State of the art and research opportunities. In: Proceedings of the international workshop on recommender systems for citizens, pp 1–6

Czerniewicz L, Agherdien N, Badenhorst J, Belluigi D, Chambers T, Chili M et al (2020) A wake-up call: equity, inequality and covid-19 emergency remote teaching and learning. Postdigit Sci Educ 2(3):946–967

DeWit A, Shaw R, Djalante R (2020) An integrated approach to sustainable development, National Resilience, and COVID-19 responses: the case of Japan. Int J Disaster Risk Reduction, 101808

Di Tommaso MR (2020a) A strategy of smart resilience for the post-coronavirus. On the role of the demand and supply of industrial policy. Industria 41(1)

Doan TT, Safavi-Naini R, Li S, Avizheh S, Fong PW (2018) Towards a resilient smart home. In: Proceedings of the 2018 workshop on IoT security and privacy, pp 15–21

Dolgikh, S. (2020). Smart-Covid: Intelligent solutions for higher risk environments. *preprint ResearchGate doi, 10.*

Erdem Ö, Derin E, Sagdic K, Yilmaz EG, Inci F (2021) Smart materials-integrated sensor technologies for COVID-19 diagnosis. Emergent Mater 1–17

Galderisi A, Colucci A (2018) Smart, resilient and transition cities: emerging approaches and tools for a climate-sensitive urban development. Elservier, Cambridge

Gil O, Cortés-Cediel ME, Cantador I (2019) Citizen participation and the rise of digital media platforms in smart governance and smart cities. Int J E-Plann Res (IJEPR) 8(1):19–34

Gupta D, Bhatt S, Gupta M, Tosun AS (2021) Future smart connected communities to fight covid-19 outbreak. Internet Things 13:100342

Inn TL (2020) Smart city technologies take on COVID-19. World Health 841

Jaiswal R, Agarwal A, Negi R (2020) Smart solution for reducing the COVID-19 risk using smart city technology. IET Smart Cities 2(2):82–88

James P, Das R, Jalosinska A, Smith L (2020) Smart cities and a data-driven response to COVID-19. Dialogues Hum Geogr 10(2):255–259

Jastrzębska AM, Vasilchenko AS (2021) Smart and sustainable nanotechnological solutions in a battle against COVID-19 and beyond: a critical review. ACS Sustain Chem Eng 9(2):601–622

Javaid M, Haleem A (2020) Exploring smart material applications for COVID-19 pandemic using 4D printing technology. J Ind Integr Manage 5(4)

Ju J, Liu L, Feng Y (2019) Public and private value in citizen participation in E-governance: evidence from a government-sponsored green commuting platform. Govern Inf Q 36(4):101400

Kang B (2021) How the COVID-19 pandemic is reshaping the education service. Future of Serv Post-COVID-19 Pandemic 1:15

Khan H, Kushwah KK, Singh S, Urkude H, Maurya MR, Sadasivuni KK (2021) Smart technologies driven approaches to tackle COVID-19 pandemic: a review. 3 Biotech 11(2):1–22

Khan MM, Karim R (2020) Development of smart e-health system for COVID-19 pandemic. In: 2020 23rd international conference on computer and information technology (ICCIT). IEEE, pp 1–6

Khatibi H, Wilkinson S, Dianat H, Baghersad M, Ghaedi K, Javanmardi A (2021) Indicators bank for smart and resilient cities: design of excellence. Built Environ Project Asset Manage. Available from: https://www.emerald.com/insight/content/doi/10.1108/BEPAM-07-2020-0122/full/html

Kher S, Nutt V, Dasgupta D (2013) Resilient hybrid overlay model for smart grid: RHM for smart grid. In: 2013 IEEE symposium on computational intelligence in cyber security (CICS). IEEE, pp 45–51

Kumar A (2017) E-governance to smart governance. 8 pages. SSRN 2970329. Available from: https://papers.ssrn.com/sol3/papers.cfm?abstract_id=2970329

Lazaroiu C, Roscia M (2018) Smart resilient city and IoT towards sustainability of Africa. In: 2018 7th international conference on renewable energy research and applications (ICRERA). IEEE, pp 1292–1298

Liu J, Guo H, Xiong J, Kato N, Zhang J, Zhang Y (2019) Smart and resilient EV charging in SDN-enhanced vehicular edge computing networks. IEEE J Sel Areas Commun 38(1):217–228

Lorente LML, Arrabal AA, Pulido-Montes C (2020) The right to education and ICT during covid-19: an international perspective. Sustainability 12(21):9091

Mcleod E, Bruton-Adams M, Förster J, Franco C, Gaines G, Gorong B et al (2019) Lessons from the Pacific islands–adapting to climate change by supporting social and ecological resilience. Front Mar Sci 6:289

Marino A, Pariso P (2020) Value governance, digital divided and economic inequality in Italy during the Covid 19 emergency. In: 2020 the 4th international conference on E-commerce, E-Business and E-Government, pp 45–53

Mezzour G, Boudanga Z, Benhadou S (2020) Smart pandemic management through a smart, resilient and flexible decision-making system. Int Arch Photogram Remote Sens Spatial Inf Sci 44:285–294

Mohammed MN, Syamsudin H, Al-Zubaidi S, AKS RR, Yusuf E (2020) Novel COVID-19 detection and diagnosis system using IOT based smart helmet. Int J Psychosoc Rehabil 24(7):2296–2303

Naresh R (2020) Education after COVID-19 crisis based on ICT tools. Purakala 31:464–468

Nguyen MH, Hargittai E, Marler W (2021) Digital inequality in communication during a time of physical distancing: the case of COVID-19. Comput Hum Behav 120:106717

Oliveira TA, Oliver M, Ramalhinho H (2020) Challenges for connecting citizens and smart cities: ICT, e-governance and blockchain. Sustainability 12(7):2926

Papa R, Galderisi A, Vigo Majello MC, Saretta E (2015) Smart and resilient cities. A systemic approach for developing cross-sectoral strategies in the face of climate change. TeMA J Land Use Mobility Environ 8 (1):19–49

Paskaleva KA (2009) Enabling the smart city: the progress of city e-governance in Europe. Int J Innov Reg Dev 1(4):405–422

Prasad YG, Maheswari M, Dixit S, Srinivasarao C, Sikka AK, Venkateswarlu B et al (2014) Smart practices & technologies for climate resilient agriculture

Riva G, Wiederhold BK, Mantovani F (2021) Surviving COVID-19: the neuroscience of smart working and distance learning. Cyberpsychol Behav Soc Netw 24 (2):79–85

Robinson L, Schulz J, Khilnani A, Ono H, Cotten SR, Mcclain N et al (2020) Digital inequalities in time of pandemic: COVID-19 exposure risk profiles and new forms of vulnerability. First Monday. Available from: https://journals.uic.edu/ojs/index.php/fm/article/view/10845

Roy T, Tariq A, Dey S (2021) A socio-technical approach for resilient connected transportation systems in smart cities. IEEE Trans Intell Transp Syst. Available from: https://ieeexplore.ieee.org/abstract/document/9312453/

Saleh Y (2020) ICT, social media and COVID-19: evidence from informal home-based business community in Kuwait City. J Enterprising Communities People Places Global Econ. Available from: https://www.emerald.com/insight/content/doi/10.1108/JEC-07-2020-0131/full/html

Sato M, Saito E (2020) How can we stop emerging educational inequality: an expert teacher's perspective on Japanese issues due to COVID-19. Manage Educ 0892020620981949

Sealy WU (2003) Empowering development through e-governance: creating smart communities in small Island States. Int Inf Libr Rev 35(2–4):335–358

Sharma PK, Park JH (2018) Blockchain based hybrid network architecture for the smart city. Futur Gener Comput Syst 86:650–655

Singh U, Rawat K, Singhla AR (2021) Dynamics of e-Governance in post COVID era: India. Electron J Inf Syst Develop Countries e12168. Available from: https://onlinelibrary.wiley.com/doi/full/10.1002/isd2.12168

Song C, Lee J (2016) Citizens' use of social media in government, perceived transparency, and trust in government. Public Perform Manag Rev 39(2):430–453

Sonn JW, Lee JK (2020) The smart city as time-space cartographer in COVID-19 control: the South Korean strategy and democratic control of surveillance technology. Eurasian Geogr Econ 61(4–5):482–492

Stoiciu A (2011) The role of e-governance in bridging the digital divide. UN Chron 48(3):37–39

Strielkowski W, Firsova I, Lukashenko I, Raudeliūnienė J, Tvaronavičienė M (2021) Effective management of energy consumption during the COVID-19 pandemic: the role of ICT solutions. Energies 14(4):893

Su Z, McDonnell D, Wen J, Kozak M, Abbas J, Šegalo S et al (2021) Mental health consequences of COVID-19 media coverage: the need for effective crisis communication practices. Glob Health 17(1):1–8

Tang T, Hou J, Fay DL, Annis C (2019) Revisit the drivers and barriers to e-governance in the mobile age: a case study on the adoption of city management mobile apps for smart urban governance. J Urban Aff 1–23

Tan-Mullins M, Cheshmehzangi A, Chien S, Xie L (2017) Smart-eco cities in China: trends and city profiles 2016. University of Exeter, Exeter (SMART-ECO Project)

Tropea M, De Rango F (2020) COVID-19 in Italy: current state, impact and ICT-based solutions. IET Smart Cities 2(2):74–81

Tsado Y, Lund D, Gamage K (2014) Resilient wireless communication networking for Smart grid BAN. In: 2014 IEEE international energy conference (ENERGYCON). IEEE, pp 846–851

Ullah A, Pinglu C, Ullah S, Abbas HSM, Khan S (2021) The role of e-governance in combating COVID-19 and promoting sustainable development: a comparative study of China and Pakistan. Chin Polit Sci Rev 6 (1):86–118

Van Fan Y, Lee CT, Lim JS, Klemeš JJ, Le PTK (2019) Cross-disciplinary approaches towards smart, resilient and sustainable circular economy. J Clean Prod 232:1482–1491

Vinod Kumar T (2015) E-governance for smart cities. In: Vinod Kumar T (ed) E-governance for smart cities. Springer, Singapore, pp 1–43

Wu G, Ono Y, Alishahi M (2015) Development of a resilient hybrid microgrid with integrated renewable power generations supplying DC and AC loads. In: 2015 IEEE international telecommunications energy conference (INTELEC). IEEE, pp 1–5

Xie L, Tan-Mullins M, Cheshmehzangi A (2019) Political ecology of Chinese smart eco-cities. In: Zhang X (ed) Remaking sustainable urbanism: space, scale, and governance in the new urban era. Palgrave Macmillan, Singapore, pp 57–78

Yang K (2020) Unprecedented challenges, familiar paradoxes: COVID-19 and governance in a new normal state of risks. Public Adm Rev 80(4):657–664

Zaman A, Islam MN, Zaki T, Hossain MS (2020) ICT intervention in the containment of the pandemic spread of COVID-19: An exploratory study. arXiv preprint arXiv:2004.09888

Zheng Y, Walsham G (2021) Inequality of what? An intersectional approach to digital inequality under Covid-19. Inf Org 31(1):100341

Zhu S, Li D, Feng H (2019) Is smart city resilient? Evidence from China. Sustain Cities Soc 50:101636

21 From "The City in Need" to "The City We Need"

Long-term plans cannot work well with short-term visions.
—Ali Cheshmehzangi, 2021

21.1 Reflecting on the Missing "Balance"

Not long ago, our discussion with a senior scholar and architect ended up debating balance in ongoing development. To our surprise, the response was rather unpleasant and perturbing. In discussing why we develop out of such balance, the reply was "we do not need the balance anymore", but we need to follow the flow. This answer was not something that we expected as we face many challenges that are usually human-made or human-triggered. For example, the calamities of health emergencies/crises to war situations and natural disasters, we are the cause of many curses, and not the cure (Cheshmehzangi and Dawodu 2019), without interrogating (adequately) that "haven't we caused enough harm?". By debating this matter, we could delve into much more significant challenges of climate change impacts, socio-economic inequalities, eating up the planet and its resources, engineering and disrupting biodiversity, manipulating eco-system services, and so on. However, we may just park these topics for future debates as we aim to focus on one of the main problems and not all.

Earlier in the first few months of the pandemic, our focus was on "The City in Need", addressing issues that mainly relate to the needs and challenges of cities and communities during the pandemics (Cheshmehzangi 2020a). As a follow-up discourse on this critical topic, this book offers a set of complementary debates on "The City We Need", focused again on facing the adversities of the pandemics. In both books, we use the COVID-19 pandemic as an excuse to study the shortfalls, shortages, and deficiencies of our cities and communities. In doing so, we reflect on challenges with an open eye, suggest ways out of better management and containment, and discuss what happened that we moved away from the city we need to the city in need.

We realise our challenges are due to the missing balance, which continuously augments in a never-ending unhealthy pattern. Our conception of growth has become a barrier to find a suitable balance-development pattern. The growing imbalance caused by centuries of power rivalries has cost us a lot in recent decades. Our advancement seems to have no visible end, and our attention to this imbalance situation has become smaller and smaller. This ongoing pandemic has taught us a lot, but as we discussed before, little will be remembered for us to reflect more considerably and responsively (Cheshmehzangi 2020b). As evidenced throughout the pandemic, we see numerous issues that suggest overlaps with our 18 factors covered in the book. In the next section, we

A. Cheshmehzangi, *Urban Health, Sustainability, and Peace in the Day the World Stopped*, Sustainable Development Goals Series, https://doi.org/10.1007/978-981-16-4888-5_21

summarise these briefly before debating the topic of health–sustainability–peace (HSP) nexus in cities and communities.

21.2 Learning from the Day the World (Almost) Stopped

From the physical dimension perspective, we see issues of unhealthy city environments affecting the most poorer and vulnerable communities. The neglect that has become a common norm in dealing with informal settlements and shanty-towns or slums worldwide is an obvious situation, which shows this is not a new issue. From the complex living conditions of Gaza in the Palestinian Territories to many unhealthy and unhygienic refugee camps around the world, we see disparities that indicate a lack of frameworks and long-term plans. In most cases, there is a lack of financial and political support mechanisms that make us think why solve the problem when it can be continuously treated in a never-ending process? In the line of the same debates, we explored issues of resilience, particularly from the morphological and spatial perspectives. We highlighted the less discussed topics of unhealthy urban landscapes, physical characteristics of cities, and socio-physical attributes in our contemporary communities. Then, we delved into critical infrastructures (CIs) challenges to discuss methods of safeguarding our city's sustainability and peace. In doing so, we highlighted many sectors under the CIs and discussed what would be relevant to do if healthy and peaceful communities are envisioned.

From the environment dimension perspective, we started looking into matters that are usually not discussed from the environmental sustainability perspective. We clearly defined differences and overlaps between environmental and ecological aspects and focused more on critical elements that are also related to our human and built (or human-built) environments. Started with debates on well-functioning government for peaceful living environments, we immediately discussed issues related to environmental governance, environmental sustainability, and related factors to the spread of the pandemics. We suggested that a well-functioning government is necessary for healthy communities, sustainable living, and peaceful environments. Afterwards, we reflected on essential systems in cities and communities, focused more on sustaining their primary operations. As a crossover with the CIs, we highlighted issues of waste management systems, water treatment systems, etc. Those are mainly caused by us, the human beings, rather than anyone or anything else. We reflected on these realities as we highlighted overarching issues of waste production and harming the environment. Afterwards, we discussed resource efficiency and distribution matters, which we believe are key drivers to healthy environments. We discussed diverse strategies that are needed for resource efficiency and reflected on the importance of self-reliant cities. Finally, we delved into issues of natural resource inefficiency and correlated the debates to natural disasters and other similar crises.

From the economic dimension perspective, we covered key areas of resilience and smartness in economies, economical operations, and business environments. A healthy business environment came up as a by-default matter in facing disruptive pandemic events. We discussed ways that businesses can foster peace and pointed out several concerns about our business environments' healthiness. During the discussions, issues of corruption, inequalities, and inequity were highlighted, too. Undoubtedly, we had to delve into destabilising city operations, caused mainly by unhealthy and unfair business environments. A silent standpoint against widespread corruption issues would have made us oppose what we meant to discuss in the book. Thus, we looked at low-level corruption as a crossover between economy and institutional structures. From international organisations to small community organisations, corruption exists and opposes peace on all scales. This inevitable human nature helped us look into monitory measures and systems, healthier partnerships and cooperation, and power-sharing debates. The follow-up investigation was on a more precise aspect of supply chain resilience in cities and communities, which have significant economic implications. In debating the topic, we looked into urban resilience and public health matters and discussed examples of supply chain networks (e.g., external and internal), sectoral and cross-sectoral issues, and opportunities for collaborative partnerships that also touch on key aspects of the sustainable development goals (SDGs).

From the social dimension perspective, we explore topics that are commonly known as big debates among scholars. One of these is related to information-sharing issues, which is widely a social matter than an institutional one. We looked at information sharing from multiple sustainability perspectives and suggested matters like resource flow, cybersecurity, and operational decisions. Our approach to information sharing entails overlapping arguments on peacebuilding mechanisms and coordination between various stakeholders, including managerial members such as governmental authorities and a larger network of partnership and societal representatives. As a follow-up, we discussed human capital and equity issues, particularly through good relationships, human capital development, and education. We explored key aspects of human capital efficiency (HCE) and related socio-economic attributes when dealing with pandemics. Partly, we discussed issues of social equity as we noted it could not be alienated from human capital discussions. In doing so, we suggested ways of further human capital development and provision of necessary training and education for the majority and not the minority. For the last aspect, we then explored more of social and public life, particularly related to other aspects such as quality of life, liveability matters, and social sustainability performance of cities and communities. The discussions again reflected on social equity and societal inequality matters.

From the institutional dimension perspective, we started our debates on public rights, justice, and fairness, particularly related to social and public services in cities. The discussions in here also highlighted matters such as social standards, social infrastructure, and human rights. Once again, we looked at structured inequalities that exist in our society and reflected on vulnerabilities of public rights, governmentalities, and public–political matters. Institutional support to public rights was discussed as we explored the institutional structures and arrangements, connectivity to external networks, and related corruption matters. Afterwards, we zoomed out to a larger scale to discuss the importance of good relations with neighbours, or good neighbourly relations. We discussed this topic as an important aspect for peacebuilding processes, ensuring that peace, power, and security are effective and for the benefit of all. Finally, we highlighted some of the critical peacebuilding and sustainability target plans and suggested more cooperation and competitiveness. By focusing beyond the city borders, we delve into shared prosperity processes in regions, such as urban–rural relations and urban–urban cooperation of the same region. The discussions were then developed further into the discourse of public networks, including multiple networks that are not just limited to social networks. In doing so, we also discussed the intersection between the public network and smart governance, which was later debated as part of the e-governance perspective from the technical solution viewpoint. We also had further discussions on multiple domains and urban systems on public networks, which then led to governance matters, institutional restructuring processes, and financial support.

From the technical dimension perspective, much of our discussions were a crossover between technical and technological matters. We noted the correlation between the two enables us to better understand technology availability, accessibility, and affordability. Once again, inequalities and inequity matters were highlighted concerning growing issues of the digital divide. While we explored examples of technological innovation, we noted the importance of multiple interventions, including those led by governments, people, and public–private partnerships. Our discussions were summarised regarding addressing issues of the digital divide, which again is a technical challenge in this growing digital age. We discussed some of the key technical–technological issues before zooming into a major aspect of asset and facilities management. This investigation highlighted the benefits of strategic asset management networks and focused more on asset-facilities integration for better urban management processes. The discussions were mostly focused on the obvious shortfall and shortages in cities and the growing demand for technical-guidance support, optimisation processes, and adaptive strategies. These suggestions led to final viewpoints on smartness and resilience matters. In this debate, we highlight more about smart-resilient models for cities and communities and a healthier future. Debating the two sides of technical solutions, we reflected on some of the ongoing debates on smart platforms, ICT-mediated platforms, and smart technologies use. Some of the discussions highlighted the role of socio-technical systems (STSs) as well as technological integration in smart-resilient practices. The suggestions were mainly focused on smart-resilient models by not relying on smart technologies but using or utilising them more smartly. To help decision-making processes, we also explored more e-governance approaches and highlighted key areas such as social media, public communication, and service delivery. Lastly, we suggested methods of achieving smart-resilient models or at least debated what directions and practices would be suitable in the future; in another emergency or crisis like the COVID-19 pandemic.

Furthermore, throughout all six dimensions, we noted the central position of balance between three key themes of health, sustainability, and peace. As sustainable development suggests a balance between nature and human well-being (Yan et al. 2018), we explore more of the governance that is needed in any development type (Book et al. 2010). From ecological and environmental balance (Kokkonen et al. 2018) to other factors such as social and public balance, we see opportunities for a better HSP nexus in cities and communities across the globe.

We also question what mattered the most? Urban health, sustainability, or peace? These are important for further debates, particularly in regard to urban development and urban sustainability research. By learning from the day the world (almost) stopped, we recognise a missing balance that for long has become the bottleneck of our sustainable living and continuity as curators. Pucci et al. (2011) discussed that human–environment relationships go back long ago, which have been mostly affected during the last two to three centuries, particularly with the rise of the Industrial Revolution in the eighteenth century. In many cases, we question our influences and interferences rather than our inspirations and interventions. In this rapidly growing innovation age, we ought to seek ways out of this growing imbalance. By out, we mean not to neglect but to resolve what needs to be resolved. Our sustainable pathways and sustainability transitions ought to become more meaningful and genuine. We need to ensure meeting the realities on the ground, rather than the imaginary support that gets us nowhere. The move towards sustainability innovation should not develop as the mean for our over-reliance on smart technologies. It should instead help us to improve balance, health, and peace in our society.

As the book focuses on a better HSP nexus in cities and communities, we hope for a new generation of more impactful instruments, policy directions, and sustainability measures. The peace that we deserve is far from reach as we struggle with overcoming the never-ending gluttony and abhorrence around the globe. If our actions turn into inactions, as some have become, we may asphyxiate society from its basic needs, further deprive the impoverished, and become more vulnerable. Instead, if we opt to become smarter, we first need to become resilient. In doing so, we have to understand the value of balance that is needed and ensure health, sustainability, and peace are achieved together. The three complement one another and interconnect so well that they could help us achieve better pathways for a more sustainable future. As we speak, our doubts remain in place, and as we progress, our concerns reproduce. To achieve a sustainable future, we must include indicators and measures that suggest healthy and peaceful processes. To accomplish these together, we may have a chance to save the future generations.

21.3 Future Research on the Health–Sustainability–Peace (HSP) Nexus

It is hard to see what the future will bring in researching the health–sustainability–peace (HSP) nexus. We hope that our today's concerns become tomorrow's witticisms and not the realities of a more imbalanced world. Nonetheless, we suggest five areas that would help to develop the HSP research further (Fig. 21.1). These are summarised below.

21.3.1 Sustainability Transitions

In this overarching area, we anticipate more joint agendas as those suggested by Kjærgård et al. (2014) for sustainability transitions, to ensure health and sustainability come together and form peaceful societies for the future. Matters like equity and prosperity should be embedded in multi-dimensional analysis, more than just environmental (World Bank 2019) or other single-dimension studies. A good example is the "handbook on sustainability transition and sustainable peace", developed by Brauch et al. (2016), which also includes the complicity of situations to achieve sustainable peace. Such approaches need to be procedural and long-term (Kates and Parris 2003), ensuring that specific transition models are generated (Brauch and Spring 2016; Brauch et al. 2016; Spring et al. 2016) for contextual requirements. Thus, we follow the guidelines that sustainability transition should be transformative sciences to ensure sustainable peace could happen (Brauch 2019). We could become more progressive and adaptive to enable new pathways of achieving HSP in all sorts of environments. A global sustainability transition may not necessarily happen due to its complexities (Weinstein et al. 2013). Still, we see opportunities that could lead to movements of HSP in the Anthropocene (Brauch et al., 2018).

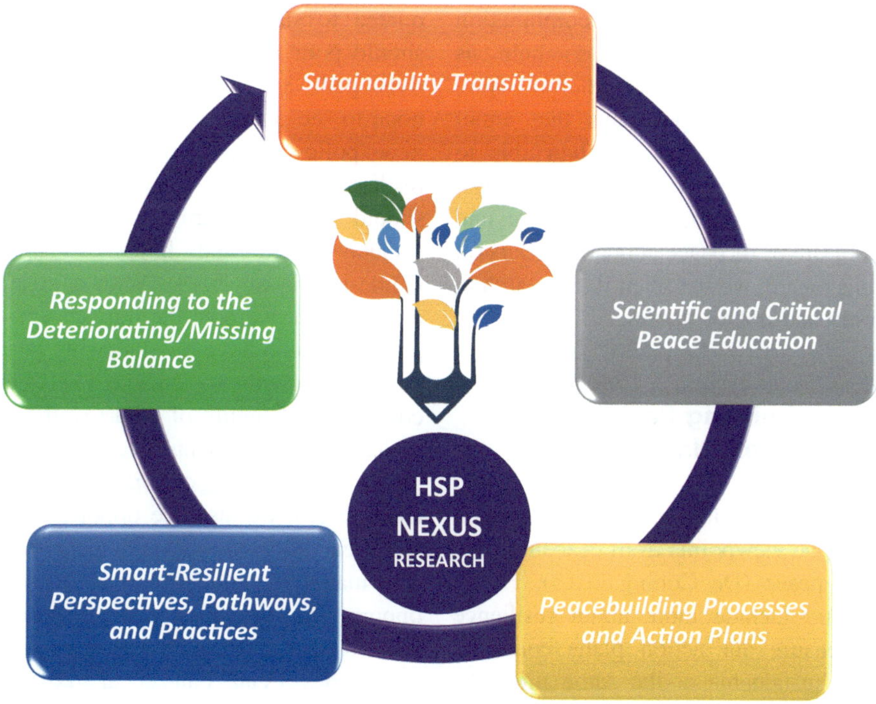

Fig. 21.1 Summary of future research on the HSP nexus. *Source* Author's own

The move towards this direction needs to be all-inclusive and reflective, or else we are stuck with long-term promises beyond our moderately short lifetime.

21.3.2 Scientific and Critical Peace Education

The need to appreciate scientific and critical peace education is evident in every sector that entails sustainability measures. As diverse as it sounds, critical peace education would be beneficial to respond to critical concerns about sustainable peace and related education (Bajaj 2008; Bajaj and Brantmeier 2011; Brantmeier 2013; Zembylas 2018). Some suggestions reflect on institutionalisation of critical peace education even in public schools, which is believed to help comprehensive implementation of peace and sustainability development through good education (Hantzopoulos 2011). By mainstreaming it, as Brantmeier (2011) suggests, we are able to "*openly embrace some of the broader and holistic purposes of education*". Its combination with scientific research would also help us develop these mainstreams even further to ensure including directions and pathways that entail health and sustainability. We are then able to move towards alternative views and pathways, which emphasis skepticism, critical thinking, and necessary objectives of peace education (Bar-Tal 2002). In this regard, we see scientific research to be a necessity and not an option.

21.3.3 Peacebuilding Processes and Action Plans

There is an urgent need to expand on research studies that argue viewpoints from peacebuilding to sustaining peace (De Coning 2016). Future research should include more about resilience thinking to ensure our action plans could be adaptive and sustainable at the same time. As Reychler and Paffenholz (2001) suggest, "*building sustainable peace is not just a matter of direct intervention through mediation*... [and]... *peacebuilders ought to monitor indicators of positive or negative developments*". We also emphasise the role of peacebuilding institutions to "*create mechanisms or a plan of action*" (True 2013) for realising the rights. The involvement of people in such processes, regardless of background, gender, ethnicity, age, and social class, means much better public participatory processes for various action plans (e.g., immediate, short-term, long-term, etc.). In such processes, we must understand differences and similarities between "*imported processes and indigenous practices*" (Millar et al. 2013), which should help us consider context-specific requirements and cultural, socio-economic, and political matters that may differ from to context. Future research must focus on practical action plans and not just peacebuilding processes.

21.3.4 Smart-Resilient Perspectives, Pathways, and Practices

An area that we hope could further develop is related to smart-resilient thinking, which then should pour new knowledge into new perspectives, pathways, and practices. Within these, we hope to see policy development or changes and new policy frameworks that include smart-resilient strategies. Cities and communities should take on board such thinking to ensure societal peace and happiness are obtained (Timashev et al. 2018). With a greater deal of appreciation, we note the impacts smart-resilient practices could have on enhancing peace and security in growing societies. It is also essential to consider co-benefits of smart-resilient thinking, which could address multiple SDGs (Mcleod et al. 2019). By leveraging opportunities for co-benefits (Kim 2018), we also see opportunities for technological interventions and technical solutions. This matter is a relatively new area where we see opportunities to exploit smart technologies (Bellini and Nesi 2018) and towards cross-disciplinary approaches (Van Fan et al. 2019). Yet, we emphasise again that smart should inform resilience, enabling for a hybrid model that promotes HSP nexus in cities and communities.

21.3.5 Responding to the Deteriorating/Missing Balance

Sometimes, we forget well-being and quality of life in the way we develop, progress, and innovate. The consideration of urban health equity (see Chap. 1) should be the backbone of decision-making processes against imbalanced urbanisation and urban development. While even in sustainability research, there is an ongoing uneasy balance (Bina 2013), we note the complexity of sustaining balance through healthy human–environment relations. Some scholars suggest a dynamic equilibrium that includes the importance of understanding multiple dimensions, such as those "*links between social capital and sustainable development*" (Dale and Onyx 2010). Thus, sustainable development strategies should aim to find the balance between multiple aspects, indicators, dimensions, and factors, especially to nurture best practices (Chaharbaghi and Willis 1999) and long-term solutions for sustainable futures. We could then see more progress from basic principles of sustainable development (Harris 2000) and towards "*maintaining system stability by developing a balance of responsibilities*" (Ivascu 2013; Duran et al. 2015). After two decades of research on this topic, mentioned first by Glasby (2002), we see the need for new paradigms and more sustainable pathways.

In all areas, we hope to see innovative interventions that would take us out of the business-as-usual (BAU) scenarios. We expect the three key themes of health, sustainability, and peace become complementary to each other, appear together more frequently in scholarly research, policies, and practices.

21.4 Towards the City We Need

It was the UN-Habitat that first developed a New Urban Agenda, with the title "The City We Need", which included an overarching idea of a new urban paradigm (UrbanLens Planning 2016). With its ten broad principles, the "cities we need" movement soon provided us with a range of global city examples that addressed diverse and sustainable pathways (see The New Urban World Journal 2017). This development soon led to a new generation of urban development strategies and approaches corresponding to the SDGs. Our discussions differ from that overarching proposal, developed after the Habitat III United Nations conference in Quito, Ecuador. The suggestions here focus more on developing a robust HSP nexus in cities and communities and a healthier and more sustainable future. We urge the immediate need to address health, sustainability, and peace together when it comes to sustainable development. We see the COVID-19 pandemic as a testing ground to identify a wide range of deficiencies and reflect on them responsibly and consider new paradigm shifts, restructuring processes, and necessary readjustments. As experienced and continue to be experienced, it was not only the people's health that was affected but also the health of cities, communities, economies, social activities, social life, institutions, etc. The impacts have been so significant that we cannot really see an end to this unfortunate calamity.

As we close the book, we reflect on key lessons gained from exploring all aspects under our six primary dimensions. The HSP nexus has been discussed throughout the book, as we hope to see more research in this area. The world we have created has become imbalanced, which shows that despite all our technological advancement and the so-called smartness, we are all vulnerable to a microscopic—but deadly—virus. This vulnerability results from the never-ending need for growth and progress, while we seem not to be fulfilled with stability and peace. This acceptance cannot happen as we seek more progress, develop without an end, and stir the imbalance even further.

From the city in need, we now talk about the city we need. We emphasise the role of the HSP nexus in achieving a more sustainable future, noting, in particular, the fact that our current efforts for sustainable development are simply not enough. For a long time, cities have become playful grounds for experimentation. Yet, we struggle to scale up best practices, fail to promote the opportunities for good education, and often choose the easiest and quickest route out of the problems without facing them or resolving them. We have lived with so many smart everything (e.g., smart cities, smart technologies, smart living, smart homes, smart cars, smart phones, etc.), but we are yet to see smartness in people. It is as if we have to revisit everything before making decisions for new paradigms and directions.

The city we need is more than just the current sustainability ideals. The more we try to learn, the more we realise we know little. As a holistic by-product, sustainable development has to promote health, sustainability, and peace—not individually but jointly and in one comprehensive package. For this book, the day the world stopped is the day the pandemic made us aware of our efficiencies in cities and communities or simply in our society. Indeed, we appraised many positive attributes throughout the book; but we see more challenges now and ahead than solutions we could ever have. Nonetheless, we believe that with a problem, there are always solutions to consider. It is just a matter of choice, as we evolve by choice and as we grow by choice. Thus, as we discussed throughout the book, we see the need for a robust HSP nexus to be embedded in our next steps of development and the future. We highlight the need for human-centric development, which will provide the backbone of a healthier and more sustainable future. The manifestation of such an approach would depend on the choices we make now, recognising our mistakes, and innovate for the better. The peace that we seek may be far from the reach, but it could happen if we see the values of a balanced life. Therefore, we have to distinguish between reality and the overindulged imaginary (or the possibility of a pseudo-reality). We have to understand the existing and imminent technologies more carefully, knowing that technologies alone will not save us. Though they will help us save the planet (and ourselves), we should opt to use them properly and smartly. In the day the world stopped, we realised the importance of health, sustainability, and peace. Yet, in many days before that and many days after, the same principles are as important as in the day the world stopped.

References

Bajaj M (2008) Critical" peace education. Encyclopedia Ppeace Educ 135–146
Bajaj M, Brantmeier EJ (2011) The politics, praxis, and possibilities of critical peace education. J Peace Educ 8(3):221–224
Bar-Tal D (2002) The elusive nature of peace education. In: Peace education: the concept, principles, and practices around the world, pp 27–36
Bellini E, Nesi P (2018) Exploiting smart technologies to build smart resilient cities. In: Gardoni P (ed) Routledge handbook of sustainable and resilient infrastructure. Routledge, New York and London, pp 685–705
Bina O (2013) The green economy and sustainable development: an uneasy balance? Eviron Plann C Gov Policy 31(6):1023–1047
Brantmeier EJ (2011) Toward mainstreaming critical peace education in US teacher education. In: Critical pedagogy in the 21st century: a new generation of scholars, pp 349–375
Brantmeier EJ (2013) Toward a critical peace education for sustainability. J Peace Educ 10(3):242–258
Book K, Eskilsson L, Khan J (2010) Governing the balance between sustainability and competitiveness in urban planning: the case of the Orestad model. Environ Policy Gov 20(6):382–396
Brauch HG (2019) Sustainable peace through sustainability transition as transformative science: a peace ecology perspective in the Anthropocene. In: Climate change, disasters, sustainability transition and peace in the Anthropocene. Springer, Cham, pp 175–234
Brauch HG, Spring ÚO, Grin J, Scheffran J (eds) (2016) Handbook on sustainability transition and sustainable peace, vol 10. Springer, Cham
Brauch HG, Spring ÚO (2016b) Sustainability transition and sustainable peace: Scientific and Policy context, Scientific concepts and dimensions. In: Handbook on sustainability transition and sustainable peace. Springer, Cham, pp 3–66
Brauch HG, Spring ÚO, Collins AE, Oswald SES (eds) (2018) Climate change, disasters, sustainability transition and peace in the Anthropocene, vol 25. Springer
Chaharbaghi K, Willis R (1999) The study and practice of sustainable development. Eng Manag J 9(1):41–48
Cheshmehzangi A (2020a) The city in need: urban resilience and city management in disruptive disease outbreak events. Springer, Singapore
Cheshmehzangi A (2020b) Reflections for 'the city in need'. In: The city in need: urban resilience and city management in disruptive disease outbreak events. Springer, Singapore, pp 285–304

Cheshmehzangi A, Dawodu A (2019) Sustainable urban development in the age of climate change—people: the cure or curse. Palgrave Macmillan, Singapore, pp 157–182
Dale A, Onyx J (eds) (2010) A dynamic balance: Social capital and sustainable community development. UBC Press, Vancouver
De Coning C (2016) From peacebuilding to sustaining peace: implications of complexity for resilience and sustainability. Resilience 4(3):166–181
Duran DC, Gogan LM, Artene A, Duran V (2015) The components of sustainable development-a possible approach. Proc Econ Finance 26:806–811
Glasby GP (2002) Sustainable development: the need for a new paradigm. Environ Dev Sustain 4(4):333–345
Hantzopoulos M (2011) Institutionalizing critical peace education in public schools: a case for comprehensive implementation. J Peace Educ 8(3):225–242
Harris JM (2000) Basic Principles of Sustainable Development, Working Paper 00-04, Global Development andEnvironment Institute, pp. 1–26, Available from: https://notendur.hi.is/~bdavids/UAU101/Readings/Harris_2000_Sustainable_development.pdf
Ivascu L (2013) Contributii privind managementul riscului in intreprinderea. Editura Politehnica, Timisoara
Kates RW, Parris TM (2003) Long-term trends and a sustainability transition. Proc Natl Acad Sci 100 (14):8062–8067
Kim KG (2018) Methods and techniques for climate resilient and low-carbon smart city planning. In: Kim KG (ed) Low-carbon smart cities. Springer, Cham, pp 177–213
Kjærgård B, Land B, Bransholm Pedersen K (2014) Health and sustainability. Health Promot Int 29 (3):558–568
Kokkonen TV, Grimmond CSB, Christen A, Oke TR, Järvi L (2018) Changes to the water balance over a century of urban development in two neighborhoods: Vancouver, Canada. Water Resour Res 54(9):6625–6642
Mcleod E, Bruton-Adams M, Förster J, Franco C, Gaines G, Gorong B et al (2019) Lessons from the Pacific islands–adapting to climate change by supporting social and ecological resilience. Front Mar Sci 6:289
Millar G, Van Der Lijn J, Verkoren W (2013) Peacebuilding plans and local reconfigurations: frictions between imported processes and indigenous practices. Int Peacekeeping 20(2):137–143
Pucci S, Pantosti D, De Martini PM, Smedile A, Munzi M, Cirelli E et al (2011) Environment–human relationships in historical times: the balance between urban development and natural forces at Leptis Magna (Libya). Quatern Int 242(1):171–184
Reychler L, Paffenholz T (eds) (2001) Peacebuilding: a field guide. Lynne Rienner Publishers, Boulder
Spring ÚO, Brauch HG, Scheffran J (2016) Sustainability transition with sustainable peace: key messages and scientific outlook. In: Handbook on sustainability transition and sustainable peace. Springer, Cham, pp 887–927
The New Urban World Journal (2017). The cities we need, vol 5(1), ISSN 2201-0203, 40 pages, Available from: https://adobeindd.com/view/publications/64090dd6-f87b-4734-b168-daa4c6c22106/o5ck/publication-web-resources/pdf/July2017NUWJ_web.pdf
Timashev SA, Alekhin VN, Poluyan LV, Fontanals I, Gheorghe A (2018) Transforming Yekaterinburg into a safe, resilient-smart and sustainable city. In: IOP conference series: earth and environmental science, vol 177, No 1, p 012001. IOP Publishing
True J (2013) Women, peace and security in post-conflict and peacebuilding contexts. NOREF policy brief. Available from: https://www.peacewomen.org/system/files/global_study_submissions/Jacqui%20True_NOREF%20policy%20brief.pdf
UrbanLens Planning (2016) The city we need 2.0: towards a new urban paradigm. Available from: https://www.urbanlensplanning.net/focalpoint/72016/pogfeoach4ahh9mhln37i570ettml0
Van Fan Y, Lee CT, Lim JS, Klemeš JJ, Le PTK (2019) Cross-disciplinary approaches towards smart, resilient and sustainable circular economy. J Clean Prod 232:1482–1491
Weinstein MP, Turner RE, Ibáñez C (2013) The global sustainability transition: it is more than changing light bulbs. Sustain Sci Pract Policy 9(1):4–15
World Bank (2019) Myanmar country environmental analysis: a road towards sustainability, peace, and prosperity—synthesis report. World Bank
Yan Y, Wang C, Quan Y, Wu G, Zhao J (2018) Urban sustainable development efficiency towards the balance between nature and human well-being: connotation, measurement, and assessment. J Clean Prod 178:67–75
Zembylas M (2018) Con-/divergences between postcolonial and critical peace education: towards pedagogies of decolonization in peace education. J Peace Educ 15 (1):1–23

Index

A. Cheshmehzangi, *Urban Health, Sustainability, and Peace in the Day the World Stopped*, Sustainable Development Goals Series, https://doi.org/10.1007/978-981-16-4888-5

D

Batch number: 10615152

Printed by Printforce, the Netherlands